GLOBAL WARMING
IN THE
21ST CENTURY

GLOBAL WARMING IN THE 21ST CENTURY

VOLUME

I   Our Evolving Climate Crisis

Bruce E. Johansen

Praeger Perspectives

Westport, Connecticut
London

Library of Congress Cataloging-in-Publication Data

Johansen, Bruce E. (Bruce Elliott), 1950–
  Global warming in the 21st century / Bruce E. Johansen.
    v.   cm.
  Includes bibliographical references and index.
  Contents: v.1. Our evolving climate crisis—v.2. Melting ice and warming seas—v.3. Plants and animals in peril
  ISBN 0-275-98585-7 (set : alk. paper)—ISBN 0-275-98586-5 (v. 1 : alk. paper)—ISBN 0-275-98587-3 (v. 2 : alk. paper)—ISBN 0-275-99093-1 (v. 3 : alk. paper)   1. Global warming.   2. Global warming—Environmental aspects.   3. Renewable energy resources. 4. Global warming—Political aspects.   I. Title: Global warming in the twenty-first century.   II. Title.
  QC981.8.G56J643   2006
  363.738'74—dc22   2006006633

British Library Cataloguing in Publication Data is available.

Library of Congress Catalog Card Number: 2006006633
ISBN: 0–275–98585–7 (set)
        0–275–98586–5 (vol.1)
        0–275–98587–3 (vol.2)
        0–275–99093–1 (vol.3)

First published in 2006

Praeger Publishers, 88 Post Road West, Westport, CT 06881
An imprint of Greenwood Publishing Group, Inc.
www.praeger.com

Printed in the United States of America

The paper used in this book complies with the Permanent Paper Standard issued by the National Information Standards Organization (Z39.48–1984).

10 9 8 7 6 5 4 3 2 1

CONTENTS

Color insert in Volume 2 precedes Part IV.

PREFACE

The Next Energy Revolution

In one hundred years, students of history may remark at the nature of the fears that stalled responses to climate change early in the twenty-first century. Skeptics of global warming kept change at bay, it may be noted, by appealing to most people's fear of change that might erode their comfort and employment security, all of which were psychologically wedded to the massive burning of fossil fuels. A necessary change in our energy base may have been stalled, they might conclude, beyond the point where climate change forced attention, comprehension, and action.

Technological change always generates fear of unemployment. Paradoxically, such changes also generate economic activity. A change in our basic energy paradigm during the twenty-first century will not cause the ruination of our economic base, as some skeptics of climate change believe, any more than the coming of the railroads in the nineteenth century ruined an economy in which the horse was the major land-based vehicle of transportation. The advent of mass automobile ownership early in the twentieth century propelled economic growth, as did the transformation of information gathering and handling via computers in the recent past. The same developments also put blacksmiths, keepers of hand-drawn accounting ledgers, and anyone who repaired manual typesetters out of work.

We are overdue for an energy system paradigm shift. Limited oil supply and its location in the volatile Middle East make a case for new sources, along with the accelerating climate change from greenhouse gases accumulating in the atmosphere. According to an editorial in *Business Week*, "A national policy that cuts fossil-fuel consumption converges with a geopolitical policy of reducing energy dependence on Middle East oil. Reducing carbon dioxide emissions is no longer just a 'green' thing. It makes business and foreign policy sense, as well. . . . In the end, the only real solution may be new energy technologies. There has been little innovation in energy since the internal combustion engine was invented in the 1860s and Thomas Edison built his first commercial electric generating plant in 1882" ("How to Combat Global Warming" 2004, 108).

Even the climate skeptics don't deny that climate has warmed. Temperatures have, indeed, been rising. Nine of the ten warmest years worldwide have been recorded since 1990. General warming does not imply that all cold weather has ended—instead, extremes have generally been increasing. As global averages have increased, for example, a spell of intense cold killed more than 1,000 people in India and Bangladesh during the winter of 2002–2003, only a few months after hundreds perished of record heat in the same area.

Increasing evidence also indicates that rising temperatures are changing the hydrological cycle, helping to cause intensifying chances of precipitation extremes of drought and deluge. Western Europe has experienced flooding rains while the western interior of North America has been suffering what may be the worst drought since that of a thousand years ago, which ruined the civilization of the Anasazis. Intensity of storms often increases with warmth. In the midst of drought during the summer of 2002, for example, sections of Nebraska experienced cloudbursts that eroded soil and washed out an interstate highway. Hours later, the drought returned.

Temperature also does not fully express itself immediately, but through a feedback loop of perhaps a half-century. We are, thus, now experiencing the climate related to greenhouse gas levels of about 1960. Since that time, carbon dioxide levels have risen substantially, all but guaranteeing further, substantial warming during at least the next half-century.

As temperatures rise, energy policy in the United States under the George W. Bush administration generally ignores atmospheric physics. Gas mileage for U.S. internal combustion engines has, in fact, declined during the last two decades, but gains in energy efficiency have been more than offset by increases in vehicle size, notably through sport-utility vehicles. And now comes a mass advertising campaign aimed at security-minded U.S. citizens for the biggest gas-guzzler of all, the fortress-like Humvee. Present-day automotive marketing may seem quaint in a hundred years. By the end of this century, perhaps sooner, the internal combustion engine and the oil (and natural gas) burning furnace will become museum pieces. They will be as antique as the horse and buggy is today. Such change will be beneficial and necessary.

As of 2005, the federal government of the United States (which, as a nation, produces almost one-quarter of the world's greenhouse gases) was sitting out the next worldwide energy revolution. The United States is being led (if that is the word) by a group of minds still set to the clock of the early twentieth-century fossil fuel boom. The Bush administration not only has refused to endorse the Kyoto Protocol, but also has (with a few exceptions, such as its endorsement of hydrogen and hybrid-fueled automobiles) failed to take seriously the coming revolution in the technology of energy production and use. In a century, George Bush's bust may sit in a greenhouse gas museum, not far from a model of an antique internal-combustion engine. A plaque may mention his family's intimate ties to the oil industry as a factor in his refusal to think outside that particular box.

As the White House banters about "sound science," yellow-jacket wasps were sighted on Northern Baffin Island during the summer of 2004. By the end of the twenty-first century, if "business as usual" fossil fuel consumption is not curbed substantially, the atmosphere's carbon dioxide level will reach 800 to 1,000 parts per million. The last time this level was reached 55 million years ago, during the days of the dinosaurs, the water at the North Pole, then devoid of ice, reached approximately 68 degrees F.

Before the end of this century, the urgency of global warming will become manifest to everyone. Solutions to our fossil fuel dilemma—solar, wind, hydrogen, and others—will evolve during this century. Within our century, necessity will compel invention. Other technologies may develop that have not, as yet, even broached the realm of present-day science fiction any more than digitized computers had in the days of the Wright Brothers a hundred years ago. We will take this journey because the changing climate, along with our own innate curiosity and creativity, will compel a changing energy paradigm.

Such change will not take place at once. A paradigm change in basic energy technology may require the better part of a century, or longer. Several technologies will evolve together. Oil-based fuels will continue to be used for purposes that require it. (Air transport comes to mind, although engineers already are working on ways to make jet engines more efficient.)

A wide variety of solutions are being pursued around the world, of which the following are only a few examples. Some changes involve localities. Already, several U.S. states are taking actions to limit carbon dioxide emissions despite a lack of support from the U.S. federal government. Building code changes have been enacted. Wind-power incentives have been enacted—even in Bush's home state of Texas, where some oil fields now host wind turbines.

Wind turbines and photovoltaic solar cells are becoming more efficient and competitive. Improvements in farming technology are reducing emissions. Deep-sea sequestration of CO_2 is proceeding in experimental form, but with concerns about this technology's effects on ocean biota. Tokyo, where a powerful urban heat island has intensified the effects of general warming, has proposed a gigantic ocean-water cooling grid. Britain and other countries are considering carbon taxes.

J. Craig Venter, the maverick scientist who compiled a human genetic map with private money, has decided to tap a $100 million research endowment he created from his stock holdings to scour the world's deep ocean trenches for bacteria that might be able to convert carbon dioxide to solid form using little sunlight or other energy. Failing that, Venter proposes to synthesize such organisms via genetic engineering. He would like to invent two synthetic microorganisms: one to consume carbon dioxide and turn it into raw materials comprising the kinds of organic chemicals that are now made from oil and natural gas, the other to generate hydrogen fuel from water and sunshine.

The coming energy revolution will engender economic growth and become an engine of wealth creation for those who realize the opportunities that it offers. Denmark, for example, is making every family a share owner in a burgeoning wind-power industry. The United Kingdom is making plans to reduce its greenhouse gas emissions 50 percent in 50 years. The British program begins to address the position of the Intergovernmental Panel on Climate Change (IPCC) that emissions will have to fall 60 to 70 percent by century's end to avoid significant warming of the lower atmosphere due to human activities. The Kyoto Protocol, with its reductions of 5 to 15 percent (depending on the country) is barely earnest money compared to the required paradigm change, which will reconstruct the system and the way most of the world's people obtain and use energy.

Solutions will combine scientific achievement and political change. We will end this century with a new energy system, one that acknowledges nature and works with its needs and cycles. Economic development will become congruent with the requirements of sustaining nature. Coming generations will be able to mitigate the effects of greenhouse gases without the increase in poverty so feared by skeptics. Within decades, a new energy paradigm will be enriching us and securing a future that works with the requirements of nature, not against it.

So, how much "wiggle room" does the Earth and its inhabitants have before global warming becomes a truly world-girdling disaster, rather than what many people take to be a set of political, economic, and scientific debating points? Perhaps the best synopsis was provided by James Hansen during a presentation at the American Geophysical Union annual meeting in San Francisco, December 6, 2005. The Earth's temperature, with rapid global warming over the past 30 years, said Hansen, is now passing through the peak level of the Holocene, a period of relatively stable climate that has existed for more than 10,000 years. Further warming of more than 1 degree C "will make the Earth warmer than it has been in a million years. 'Business-as-usual' scenarios, with fossil fuel CO_2 emissions continuing to increase at about 2 percent a year as in the past decade, yield additional warming of 2 or 3 degrees C this century and imply changes that constitute practically a different planet" (Hansen 2005).

Stop for a moment, and ponder the words, delivered in the measured tones of a veteran scientist: "Practically a different planet," a very real probability by the end of the twenty-first century. Hansen is not joking. He continued: "I present multiple lines of evidence indicating that the Earth's climate is nearing, but has not passed, a tipping point, beyond which it will be impossible to avoid climate change with far-ranging undesirable consequences" (Hansen 2005).

Coming to cases, Hansen described changes that will include:

not only loss of the Arctic as we know it, with all that implies for wildlife and indigenous peoples, but losses on a much vaster scale due to worldwide rising seas. Sea level will increase slowly at first, as losses at the fringes of Greenland and Antarctica due to accel-erating ice streams are nearly balanced by increased snowfall and ice sheet thickening in the ice sheet interiors. But as Greenland and West Antarctic ice is softened and lubricated by melt-water and as buttressing ice shelves disappear due to a warming ocean, the balance will tip toward ice loss, thus bringing multiple positive feedbacks into play and causing rapid ice sheet disintegration. The Earth's history suggests that with warming of 2 to 3 degrees C the new equilibrium sea level will include not only most of the ice from Greenland and West Antarctica, but a portion of East Ant-arctica, raising sea level of the order of 25 meters. (80 feet)

To be judicious—we don't want to ruin our case with over-statement—one might allow perhaps two or three centuries for a tem-perature rise in the atmosphere to express itself as sea-level rise from melting ice. Contrary to lethargic ice sheet models, Hansen suggests, real-world data suggest substantial ice sheet and sea-level change in centuries, not millennia. Now take a look at a map of the world and pay attention to the coastal urban areas. Is anyone worried yet?

Hansen hopes that "the grim 'business-as-usual' climate change" may be avoided by slowing the growth of greenhouse gas emissions during the first quarter of the present century, requiring "strong policy leadership and international cooperation" (Hansen 2005). However, he noted (venturing into the realm of politics) that "special interests have been a roadblock wielding undue influence over policymakers. The special interests seek to maintain short-term profits with little regard to either the long-term impact on the planet that will be inherited by our children and grandchildren or the long-term economic well-being of our country" (Hansen 2005). Hansen leaves to the audience the task of putting names and faces to the special interests who, along with the rest of us, are attending this crucial juncture in the history of the planet and its inhabitants.

FURTHER READING

Hansen, James E. Is There Still Time to Avoid "Dangerous Anthropogenic Interference" with Global Climate? A Tribute to Charles David Keeling. Paper delivered to the American Geophysical Union, San Francisco, December 6, 2005. www.columbia.edu/~jeh1/keeling_talk_and_slides.pdf.

"How to Combat Global Warming: In the End, the Only Real Solution May Be New Energy Technologies," *Business Week*, August 16, 2004, 108.

ACKNOWLEDGMENTS

Anyone who has written and published a book knows well that it is hardly a solitary journey, even after many hundreds of hours alone at the keyboard. Along the way, many thanks are due, in my case to my wife Pat Keiffer and my family (Shannon, Samantha, Madison), who kept me clothed and fed while enduring numerous bulletins from global warming's many scientific and political fronts. Gratitude also is due to the people of University of Nebraska at Omaha Interlibrary Loan, who can get just about anything that's been published anywhere; to my editors Heather Staines and Lisa Pierce; to the production crew; to the University of Nebraska at Omaha School of Communication Director Jeremy Lipschultz (himself an accomplished author who knows my working habits); and Deans Robert Welk and Shelton Hendricks (for partial relief from teaching duties). Further debts are owed to manuscript reviewers Andrew Lacis of NASA's Goddard Institute for Space Studies in New York City; Gian-Reto Walther of the Institute of Geobotany, University of Hannover, Germany; and Julienne Stroeve, with the Cooperative Institute for Research in Environmental Studies, University of Colorado, Boulder.

INTRODUCTION: VOLUME 1

Global Warming as a Weapon of Mass Destruction

For fifteen years now, some small percentage of the world's scientists and diplomats have inhabited one of those strange dreams where the dreamer desperately needs to warn someone about something bad and imminent but, somehow, no matter how hard he shouts, the other person in the dream . . . can't hear him . . . . The world is about to change more profoundly than at any time in the history of human civilization.

—Bill McKibben

Lord Peter Levene, board chairman of Lloyd's of London, has said that terrorism is not the insurance industry's biggest worry, despite the fact that his company was the largest single insurer of the World Trade Center. Levene said that Lloyd's, along with other large international insurance companies, is bracing for an increase in weather disasters related to global warming (Newkirk 2003, 3-D). During January 2005, Rajendra Pachauri, chairman of the Intergovernmental Panel on Climate Change (IPCC), said, with regard to global warming, "We are risking the ability of the human race to survive" (Hertsgaard 2005).

Levene made these remarks late in 2002. A month after Hurricane Katrina hit the U.S. Gulf of Mexico coast in late summer 2005, many insurance executives expected the industry's cost to be about $35 billion, which equaled the inflation-adjusted private payout for the World Trade Center attacks in 2001. The previous U.S. industry record for a season of U.S. hurricanes had been about $27 billion in 2004, when four powerful hurricanes hit Florida. The pre-Katrina hurricane record payout (also inflation-adjusted) was $23 million, for Hurricane Andrew, during 1992 (Treaster 2005). Total insurance losses in the United States due to damage from hurricanes Katrina, Rita, and Wilma amounted to $57.6 billion by the end of 2005, more than twice 2004's previous all-time record. During 2005, world insurance loses due to weather-related natural disasters estimated by the Munich Re Foundation, part of one of the world's largest reinsurance companies, rose to an all-time record of more than U.S. $200 billion.

Following his assignment as chief weapons inspector in Iraq, Hans Blix said: "To me the question of the environment is more ominous than that of peace and war. We will have regional conflicts and use of force, but world conflicts I do not believe will happen any longer. But the environment, that is a creeping danger. I'm more worried about global warming than I am of any major military conflict" ("Hans Blix's Greatest Fear" 2003, D-2). Sir John Houghton, cochair of the Intergovernmental Panel on Climate Change, agreed. "Global warming is already upon us," he said. "The impacts of global warming are such that that I have no hesitation in describing it as a weapon of mass destruction" (Kambayashi 2003, A-17). Speaking at the Aspen Ideas Festival in July 2005, Bill Clinton said that the greatest threat facing mankind this century is not nuclear, biological, or chemical terrorism, but global warming ("Does Oil" 2005, 32).

Mark Lynas, an author who has written extensively on global warming, traveled around the world cataloguing impacts of climate change for a book, *High Tide* (Lynas 2004), said, "This is a global emergency. We are heading for disaster and yet the world is still on fossil fuel autopilot. There needs to be an immediate phase-out of coal, oil and gas, and a phase-in of clean energy sources." According to Lynas: "People can no longer ignore this looming catastrophe" (Reynolds 2003, 6).

Author Ross Gelbspan agreed: "Climate change is not just another issue. It is *the* overriding threat facing human civilization in the twenty-first century, and so far our institutions are doing dangerously little to address it. Americans in particular are still in denial, thanks largely to the efforts of the fossil fuel industry and its allies in the Bush administration. But the nation's biggest environmental organizations and opposition politicians have also displayed a disturbing lack of leadership on this crucial challenge" (Gelbspan 2004, 24).

Weather is the story; climate is the plot. The innate variability of daily weather sometimes masks longer-term trends. Climatic news comes to us in snapshots: Thousands die of heat in Europe. Corals bleach under heat stress. Polar bears go hungry, lacking seasonal ice from which to hunt seals.

In *Global Warming in the 21st Century*, I attempt to describe the science of global warming in terms that a nonspecialist reader can understand. As if I would ever get the chance, I might even pretend that I am addressing George W. Bush's knitted brow. I also want to lay out present-day evidence that we already have started down the climatic path to a state of geophysical affairs in the atmosphere, and therefore on the surface of the Earth, that will make generations to come very uncomfortable.

Chapters 1 through 5 sketch the evolutionary nature of global-warming science. Chapter 1 concentrates on the history of the idea and some basic concepts. Controversies and questions in contemporary climate-change science are developed in Chapters 2 through 4. Special attention is reserved in Chapter 3 for feedbacks that will accelerate the pace of warming later in the century. Chapter 4 is devoted to the relationship between global warming near the surface of the Earth and depletion of ozone in the stratosphere.

These chapters attempts to make understandable (and salient) to interested nonscientists the unknowns of the carbon cycle, the "stories" told by ice-core sampling, questions regarding how much of a change in carbon dioxide levels might produce a substantial change in temperatures, and how quickly this might happen.

Some recent science has undermined old assumptions. One such assumption regards the role of forests in carbon sequestration; some evidence now suggests that lush grasslands might remove more carbon dioxide from the atmosphere than most forests. Knowledge evolves on a steady basis on such subjects as the role of aerosols, including black soot, in climate change, the speed with which warming might take place, and other important questions.

Understanding of feedbacks is crucial to an understanding of why many scientists believe that the severity of global warming will become much more serious with each passing decade of the twenty-first century. These feedbacks include release of carbon dioxide and methane from arctic peat and underwater methane clathrates (solid deposits in the oceans) as well as release of carbon via wildfires.

Geophysical evidence suggests that the Earth has suffered bouts of severe warming in the distant past from natural causes that were intensified by release of the planet's stores of greenhouse gases. Considerable scientific inquiry is now aimed at estimating just how much human-provoked warming might cause the process to reach a "runaway" status in which the feedbacks take control and force warming out of control.

Several effects of warming compound each other in synthesis. For example, shrinking snow cover, with its high reflectivity, allows polar surfaces to absorb more heat on sea and land. The warming of land surfaces melts permafrost, which releases larger amounts of carbon dioxide and methane. The cycle reinforces itself.

Researchers have reported that Earth's ancient stores of peat are gasifying into the atmosphere at an accelerating rate that is adding significantly to the atmosphere's overload of greenhouse gases. Given the fact that one-third of the Earth's carbon is stored in far northern latitudes (mainly in tundra and boreal forests), the speed with which warming of the ecosystem releases this carbon dioxide to the atmosphere is vitally important to forecasts of global warming's speed and effects. The amount of carbon stored in arctic ecosystems also comprises two-thirds of the amount presently found in the atmosphere. Its release into the atmosphere will depend upon the pace of temperature rise—and the Arctic, according to several sources, has been the most rapidly warming region of the Earth.

Perhaps the most important practical implication of feedback loops is their delayed effect. Because of feedbacks, global warming is a pernicious, slow-motion crisis to which debate and diplomacy in human societies is uniquely ill-suited. Yesterday's sport-utility vehicle exhaust does not instantly become tomorrow's rising temperature. Through an intricate set of feedback loops, fossil fuel burned today is expressed in warming forty to fifty or more years later. Today we are seeing temperatures related to fossil fuel emissions from roughly 1960, when worldwide fossil fuel consumption was much lower what it was in 2004. Today's fossil fuel emissions will be expressed in the atmosphere about 2050. The implications of feedback loops are enormous; in the unlikely event that human beings were to cut their emissions of greenhouse gases to zero tomorrow, the atmosphere near the Earth's surface still would continue to warm for several decades. Temperatures in the oceans would continue to rise for at least a century, probably longer. The oceans would continue to warm and ice would continue to melt at least as long.

A monitoring station on the summit of Hawaii's Mauna Loa has been tracking increases in the level of atmospheric carbon dioxide during the past fifty years. These readings indicate sharp increases in the rate at which the greenhouse gas has been accumulating in the atmosphere. The recent increases—2.08 parts per million (ppm) from 2001 to 2002 and 2.54 ppm from 2002 to 2003—have drawn attention of climate scientists because they deviate from an historic average annual increases of around 1.5 ppm.

A debate has arisen: Are these increases an aberration or evidence of an accelerating rate of carbon dioxide buildup? Is this accelerating rate of increase the first evidence of a "runaway greenhouse effect" stoked by a series of feedback mechanisms that will cause worldwide temperatures to rise at a much more rapid rate, along with accelerating changes of climate, melting ice caps, and quickly rising sea levels?

Another important source for carbon dioxide has been provided by Indonesian fires that polluted air over Southeast Asia during the El Niño years of 1997 and 1998. An area twice the size of Belgium burned in Indonesia during 1997. Susan Page of Britain's University of Leicester, together with colleagues in England, Germany, and Indonesia, analyzed satellite photos and data gathered on the ground to estimate how much of the fire area's living vegetation and peat deposits burned.

In Indonesia, layers of peat as thick as twenty meters (sixty-six feet) cover an area of about 180,000 square kilometers (112,000 square miles) in Kalimantan (Borneo), Sumatra, and Papua New Guinea. Page and colleagues used satellite images of a 2.5-million-hectare study area in central Kalimantan from before and after the 1997 fires. According to their estimates, about 32 percent of the area had burned, of which peat land accounted for 91.5 percent. An estimated 0.19 to 0.23 gigatons of carbon were released to the atmosphere through peat combustion, with a further 0.05 gigaton released from burning of the overlying vegetation. Extrapolating these estimates to Indonesia as a whole, the researchers estimated that between 0.81 and 2.57 gigatons of carbon were released to the atmosphere in 1997 as a result of burning peat and vegetation in Indonesia.

Page and colleagues reported in *Nature* that the carbon dioxide released by these fires was "equivalent to 13 to 40 percent of the mean annual global carbon emissions from fossil fuels, and contributed greatly to the largest annual increase in atmospheric CO_2 concentration detected since records began in 1957" (Page et al. 2002, 61).

The role of ozone depletion in the world's greenhouse equation is considered in Chapter 4. When chlorofluorocarbons (CFCs) were banned in the late 1980s, most experts expected ozone depletion over the Antarctic to heal. By 2005, ozone depletion there still was a major problem, and an ozone "hole" was beginning to open over the Arctic as well. The nature of science has evolved during those fifteen years to explain how the capture of heat near the Earth's surface by greenhouse gases speeds cooling in the stratosphere and plays an important role in continuing ozone depletion at that level. Thus, the healing of the stratospheric ozone layer depends, to some degree, on reduction of greenhouse gas levels in the lower atmosphere.

Chapter 5, "Weather Wars: Global Warming and Public Opinion," examines the politics of climate change, from the attitudes of George W. Bush's presidency to China and India's roles as "wild cards" in the future. Cultural artifacts of our time (such as the movie *The Day After Tomorrow* and Michael Crichton's novel *State of Fear*) are analyzed vis-à-vis the broader political debate over global warming and the enormous economic stakes of greenhouse gas control.

Following its survey of scientific issues, the first volume of *Global Warming in the 21st Century* assesses weather and climate change as most people observe it with four chapters (6 through 9). This section begins in the present tense, as persistent warmth has been noted worldwide, with a few exceptions. These chapters gather anecdotal evidence from many sources worldwide in an attempt to sketch one large portrait of our world's changing climate.

The evidence does not always flow in one direction. Climate skeptics, for example, will long remember the white Christmas of 2004 in Brownsville, Texas. Someone even sold a Brownsville snowball on eBay for $92. Chapter 6 describes how human influences now dominate climate change as temperatures rise. Could Europe's searing heat wave of 2003, which killed more than 30,000 people, become average climatic fare by the end of the century?

Chapter 7 concentrates on ways in which a warming climate intensifies the hydrological cycle, paradoxically intensifying both drought and deluge—sometimes, alternately, in the same location. That chapter surveys the science of the changing hydrological cycle, and it ends with consideration of the debate regarding whether a warming climate might cause hurricanes to intensify as well as its impact on the worldwide spread of deserts, a prospective increase in the number of "environmental refugees," and their impact on disaster relief.

Warmer air holds more moisture, generally making rain (and sometimes snow) heavier. Warmer air also increases evaporation, paradoxically intensifying drought at the same time. While models of a warming climate generally agree that atmospheric moisture will increase with temperature, theory as well as an increasing number of daily weather reports indicates strongly that increases in precipitation will not be evenly distributed across time and space. They will, in fact, be highly uneven, episodic, and sometimes damaging. Both drought and deluge are likely to become more severe.

By 2000, the hydrological cycle seemed to be changing more quickly than temperatures. In 2005, for example, India's annual monsoon brought a thirty-seven-inch rainfall in twenty-four hours to Mumbai

(Bombay). India is accustomed to a drought-and-deluge cycle, but not like this. With sustained warming, usually wet places generally seemed to be receiving more rain and snow than before; dry places often received less precipitation and became subject to more persistent drought.

Atmospheric moisture increases rapidly as temperatures rise; over the United States and Europe, atmospheric moisture increased 10 to 20 percent from 1980 to 2000. "That's why you see the impact of global warming mostly in intense storms and flooding like we have seen in Europe," Kevin Trenberth, a scientist with the National Center for Atmospheric Research, told London's *Financial Times* (Wolf 2000, 27; see also Trenberth 2003, 1205–1217). As if on cue to support climate models, the summer of 2002 featured a number of climatic extremes, especially regarding precipitation. Excessive rain deluged Europe and Asia, swamping cities and villages and killing at least 2,000 people, while drought and heat scorched much of the United States. Climate skeptics argued that weather is always variable, but other observers noted that extremes seemed to be more frequent than before.

Thomas Karl, director of the National Climatic Data Center in the United States government's National Oceanic and Atmospheric Administration, said, "It is likely that the frequency of heavy and extreme precipitation events has increased as global temperatures have risen. This is particularly evident in areas where precipitation has increased, primarily in the mid and high latitudes of the Northern Hemisphere" (Hume 2003, A-13). Studies at the Goddard Institute for Space Studies and Columbia University indicate that the frequency of heavy downpours has indeed increased and suggest that trend will intensify.

Chapter 8 brings the climate change focus to the United States and Canada, including regional reports on warming: the prospective demise of the maple syrup industry in New England, the relationship between warming climate and increasing numbers of wildfires, the effects of shrinking snow pack in the mountains of western North America, and other matters. Chapter 9 considers similar case studies around the world, mainly on the British Isles, where climate change has received major political attention since the days of Prime Minister Margaret Thatcher, a former chemistry teacher.

These are some of the reasons why the insurance industry worries about the weather. It makes book on financial risk in the future, and what's coming this century will be costly to life and property around the world. *Global Warming in the 21st Century* is offered in an attempt to explain why many climate scientists consider global warming an urgent, worldwide problem requiring a high degree of political attention and resolve.

FURTHER READING

"Does Oil Have a Future?" (Editorial). *Atlantic Monthly*, October 2005, 31–32.

Gelbspan, Ross. "Boiling Point." *The Nation*, August 16, 2004, 24–27.

"Hans Blix's Greatest Fear." *New York Times*, March 16, 2003, D-2.

Hertsgaard, Mark. "It's Much Too Late to Sweat Global Warming." *San Francisco Chronicle*, February 13, 2005.

Hume, Stephen. "A Risk We Can't Afford: The Summer of Fire and the Winter of the Deluge Should Prove to the Nay-Sayers That If We Wait Too Long to React to Climate Change We'll be in Grave Peril." *Vancouver Sun* (Canada), October 23, 2003, A-13.

Kambayashi, Takehiko. "World Weather Prompts New Look at Kyoto." *Washington Times*, September 5, 2003, A-17.

Lynas, Mark. *High Tide: News from a Warming World*. London: Flamingo, 2004.

McKibben, Bill. "Worried? Us?" *Granta* 83 (Fall 2003): 7–12.

Newkirk, Margaret. "Lloyd's Chief Sees No Relief in Premiums; Insurance Firms Rebuild Reserves." *Atlanta Journal-Constitution*, October 21, 2003, 3-D.

Page, Susan E., Florian Siegert, John O. Rieley, Hans-Dieter V. Boehm, Adi Jaya, et al. "The Amount of Carbon Released from Peat and Forest Fires in Indonesia during 1997." *Nature* 420 (November 7, 2002): 61–65.

Reynolds, James. "Earth is Heading for Mass Extinction in Just a Century." *The Scotsman*, June 18, 2003, 6.

Treaster, Jospeh B. "Gulf Coast Insurance Expected to Soar." *New York Times*, September 24, 2005. www.nytimes.com/2005/09/24/business/24insure.html.

Trenberth, Kevin E., Aiguo Dai, Roy M. Rassmussen, and David B. Parsons. "The Changing Character of Precipitation." *Bulletin of the American Meteorological Society*, September 2003, 1205–1217.

Wolf, Martin. "Hot Air about Global Warming." *Financial Times* (London), November 29, 2000, 27.

I GLOBAL WARMING SCIENCE

The Evolving Paradigm

INTRODUCTION

For the past two centuries, at an accelerating rate, the basic composition of the Earth's atmosphere has been materially altered by the fossil fuel effluvia of machine culture. Human-induced warming of the Earth's climate is emerging as one of the major scientific, social, and economic issues of the twenty-first century, as the effects of climate change become evident in everyday life in locations as varied as small island nations of the Pacific Ocean and the shores of the Arctic Ocean. During June 2004, scientists detected rapid growth in airborne concentrations of carbon dioxide. Carbon dioxide levels recorded during March 2004 at Hawaii measured 379 parts per million (ppm), an increase of 3 ppm compared to the previous year. By comparison, there had been an annual increase averaging 1.8 ppm during the previous decade. The increases in 2004 followed increases of 2.08 and 2.54 ppm in 2002 and 2003 respectively, igniting speculation among scientists that atmospheric carbon dioxide might be reaching a runaway ascent (Keeling and Whorf 2004). Before 2002, a back-to-back reading of 2.0 ppm or more had never been recorded, and the only other years with increases of 2.0 or more (1973, 1988, 1994, and 1998) had involved El Niño conditions. No El Niño conditions were noted in 2003 or 2004. The Mauna Loa records shows an 18.8 percent increase in the mean annual concentration, from 315.98 ppm by volume of dry air in 1959 to 375.64

ppm in 2003. The El Niño–aided 1997–1998 increase of 2.87 ppm represented the largest single yearly jump since the Mauna Loa records began in 1958 (Keeling and Whorf 2004).

"The risks of global warming are real, palpable, the effects are accumulating daily, and the costs of correcting the trend rise with each day's delay," warned George M. Woodwell, director of the Woods Hole Research Center ("Eco Bridge" n.d.). Dean Edwin Abrahamson, an early leader in the field, commented: "Fossil fuel burning, deforestation, and the release of industrial chemicals are rapidly heating the earth to temperatures not experienced in human memory. Limiting global heating and climatic change is the central environmental challenge of our time" (Abrahamson 1989, xi).

This section considers scientific issues related to global warming, including various debates about the carbon cycle, a natural mechanism that is still being discovered. Are forests, for example, overrated as a carbon sink? How is global warming related to drought cycles and monsoons? How important a role is played by soot? How quickly might climate change occur in the coming century? How does land use factor into the overall picture? What feedback loops (such as increasing water vapor, gasification of permafrost, and the release of solid methane from the ocean floor) might enhance warming from greenhouse gas emissions? How is global warming near the surface of the Earth related to ozone depletion in the stratosphere?

At the end of this scientific survey, we return to the "weather wars" with a contemporary description of public opinion and climate change.

1 CLIMATE-CHANGE SCIENCE: THE BASICS

Alarm bells have been ringing regarding global warming in the scientific community for the better part of two decades. A statement issued in Toronto during June 1988, representing the views of more than 300 policy makers and scientists from forty-six countries, the United Nations, and other international organizations, warned:

> Humanity is conducting an unintended, uncontrolled, globally pervasive experiment whose ultimate consequences could be second only to nuclear war. The earth's atmosphere is being changed at an unprecedented rate by pollutants resulting from human activities, inefficient and wasteful fossil fuel use and the effects of rapid population growth in many regions. These changes are already having harmful consequences over many parts of the globe. (Abrahamson 1989, 3)

Michael Meacher, speaking as Great Britain's environment minister, has said, "Combating climate change is the greatest challenge of human history" (Brown 1999, 44). If the atmosphere's carbon dioxide level doubles over preindustrial levels, which is likely (at present rates of increase) during the twenty-first century, several climate models indicate that temperatures might rise 1.9 to 5.2 degrees C (3.4 to 9.4 degrees F), producing "a climate warmer than any in human history. The consequences of this amount of warming are unknown and could include extremely unpleasant surprises" (National Academy of Sciences 1991, 2).

The problem is at once very simple and also astoundingly complex. Increasing human populations, rising affluence, and continued dependence on energy derived from fossil fuels are the crux of the issue. The complexity of the problem is illustrated by the degree to which the daily lives of industrial-age peoples depend on fossil fuels. This dependence gives rise to an array of local, regional, and national economic interests. These interests cause tensions between nations attending negotiations to reduce greenhouse gas emissions. The cacophony of debate also illustrates the strength and diversity of established interests that are being assiduously protected. Add to the human elements of the problem the sheer randomness of climate (as well as the amount of time that passes before a given level of greenhouse gases is actually factored into climate), and the problem has become complex and intractable enough to (thus far) seriously impede any serious, unified effort by humankind to fashion solutions.

The "greenhouse effect" is not an idea that is new to science. It merely has become more easily detectable in our time as temperatures have risen and scientists have devised more sophisticated ways to measure and forecast atmospheric processes. The atmospheric balance of trace gases actually started to change beyond natural bounds at the dawn of the industrial age, with the first large-scale burning of fossil fuels. The greenhouse effect first became noticeable in the 1880s. After an intensifying debate, the idea that human activity is warming the earth in potentially damaging ways became generally accepted in scientific circles by about 1995.

Taken to extremes, an atmosphere beset by the greenhouse effect can be very unpleasant—witness perpetually cloudy Venus, with an atmosphere that is 96 percent carbon dioxide. The surface temperature on Venus, heated considerably by runaway greenhouse warming, is roughly 840 degrees F, hot enough to melt lead. The planet Mars' atmosphere is 95 percent carbon dioxide, but it's so thin that temperatures on the surface average minus 53 degrees C. Earth's moderate degree of infrared

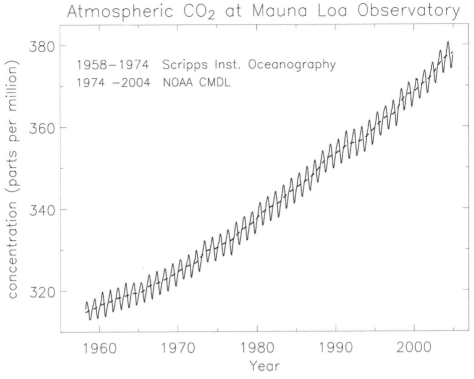

Atmospheric carbon dioxide levels as measured at the Manua Loa observatory, Hawaii, 1957–2004. Courtesy of the National Oceanic and Atmospheric Administration.

forcing (along with its blanket of liquid water) keep the planet habitable.

RISING LEVELS OF GREENHOUSE GASES
DUE TO FOSSIL FUEL EFFLUVIA

Since the beginning of the industrial age, roughly three centuries ago, an increasing world human population has made more widespread use of fossil fuels to aid economic development as well as to augment human comfort, convenience, and financial profit. Combustion of oil, coal, and natural gas has been changing the atmospheric balance of carbon dioxide, methane, nitrous oxides, and other naturally occurring trace gases as well as chemicals created by industry, such as chlorofluorocarbons (CFCs). Detectable increases in most of these gases, all of which retain heat in the atmosphere, can be traced to the beginning of the nineteenth century. The rise in greenhouse gas levels was small at the time, with effects that were very difficult to separate from the natural variability of climates.

The Earth's atmosphere is composed of 78.1 percent nitrogen and 20.9 percent oxygen. All the other gases, including those responsible for the greenhouse effect, make up only about 1 percent of the atmosphere. Carbon dioxide (CO_2) is 0.035 percent, methane (CH_4) is 0.00017 percent, and ozone is 0.000001 to 0.000004 percent. The greenhouse effect is absolutely necessary to keep the Earth at a temperature that sustains life as we know it. Like all good things in life, however, having too much can be as bad as having too little. Without the greenhouse effect, the average temperature of the Earth would be about 33 degrees C (60 degrees F)—colder than today's averages, too cold to sustain the Earth's existing plant and animal life.

During 1860, human-induced carbon emissions stood at about one-tenth of a gigaton (billion metric tons) per year. Between 1900 and 1940, human carbon production rose from roughly 1.0 to 1.5 gigatons each year. During the 1940s, yearly usage began to increase more rapidly, passing 3.0 gigatons about 1960, 5.0 about 1970, and more than 8.0 gigatons by the late 1980s. Between 1950 and 1980, worldwide emissions of carbon dioxide increased 219 percent, an average of 7.3 percent a year. The rate of increase slowed during the 1980s, but the world total still grew almost 13 percent between 1980 and 1988 (Kane and South 1991, 193). Between 1850 and 2000, human combustion of fossil fuels increased fiftyfold. Contemporary observations indicate that the carbon dioxide content of the atmosphere is increasing at between 1.0 and 3.5 parts per million (ppm) each year, with large annual variations. In some years, such as 1972 and 1988, atmospheric carbon dioxide increased 2.5 to 3.0 percent, while during one year, 1973, it did not increase at all. Roughly two-thirds of the increase in carbon dioxide since industrialism began has taken place since 1950.

Carbon dioxide is only the best known of several gases that contribute to global warming, but it is responsible for about half of the greenhouse effect. Several other gases, the most notable being methane, contribute to the other half. Water vapor also plays a role, but as a feed-back mechanism rather than a greenhouse gas. As the air warms, it holds more water vapor. The atmosphere holds 6 percent more water vapor with each 1 degree C rise in temperature.

Today, mainly as a result of the combustion of fossil fuels, the balance of heat-retaining gases in the atmosphere is increasing rapidly. In a century-and-a-half of rapid worldwide industrialization, the proportion of carbon dioxide has risen from roughly 280 to about 380 ppm. By the year 2004, scientists had tested proxies for the composition of the at-mosphere to roughly 60 million years in the past. The level of carbon dioxide today is believed, according to such measurements, to be as high as it has been in at least 20 million years.

The levels of carbon dioxide, methane, and other greenhouse gases in the atmosphere during the last years of the twentieth century was higher than at any time since humankind has walked the Earth, a circumstance that has provoked one student of the greenhouse effect to remark, "We are headed for rates of temperature rise unprecedented in human history; the geological record screams a warning to us of just how unprecedented [the stresses on] the natural environment will be" (Leggett 1990). The U.N. General Assembly's Intergovernmental Panel on Climate Change's *Second Assessment* (1995) found that the average temperature for the period 1901–1990 was higher than for any ninety-year interval since at least AD 914. The panel also has stated that the combined effect of all greenhouse gases is likely to produce a warming greater, in terms of its speed, than any other climatic event in the last 10,000 years.

GREENHOUSE GASES AND EVERYDAY LIFE

Emissions of carbon dioxide and other greenhouse gases are built into our everyday lives—coming from our modes of transportation, pro-duction, and consumption—to the extent that they are enabled by the combustion of fossil fuels. Roughly 80 percent of human industrial ac-tivity worldwide is fueled by the combustion of energy that produces carbon dioxide (and, oftentimes, other greenhouse gases as well). The same industrial processes also produce waste heat in addition to greenhouse gases. Sometimes these manufactured goods (such as automobiles) also produce waste heat and greenhouse gases as they are operated.

In addition to carbon dioxide, during the 1990s, human activity was adding about 550 metric tons of methane to the atmosphere each year. Methane's preindustrial range in the atmosphere was 320 to 780 parts per billion (ppb); by 2004, that level had risen to more than 1,700 ppb, a steeper rise, in proportional terms, than carbon dioxide's. Carbon dioxide is 200 times more plentiful in the atmosphere than methane, but a molecule of methane can trap more than ten times as much heat as one of carbon dioxide.

Atmospheric methane is produced by many human activities, from transporting natural gas to the raising of meat animals, the dumping of garbage in landfills, and the growing of rice. Methane contributes about half as much retained heat to the atmosphere as carbon dioxide. The rate of increase in atmospheric methane (1 percent a year) was about twice as rapid as that of carbon dioxide for much of the twentieth century (Jager and Ferguson 1991, 79).

Carbon monoxide shares greenhouse gas properties with carbon dioxide; its presence in the atmosphere has been rising 0.8 to 1.5 percent a year. Most of the carbon monoxide in the atmosphere is produced by humans, much of it by the burning of fossil fuels and deforestation. By the late 1980s, roughly 1 billion metric tons of human-produced carbon monoxide was being added to the global atmospheric inventory each year.

Add to this mix of trace gases tropospheric ozone, which is produced photochemically in the atmosphere from the oxidation of carbon monoxide, methane, or other hydrocarbons in the presence of nitrogen oxides, which act as catalysts. Tropospheric ozone was increasing about 1 percent a year by the late 1980s. Levels in the air have increased 20 to 50 percent during the twentieth century. This type of ozone, which is contributed to the atmosphere by several industrial processes, acts to absorb infrared radiation and, thereby, like carbon dioxide and methane, contributes to the greenhouse effect.

A few other gases, such as nitrous oxide (known as "laughing gas"), add to the greenhouse effect as well. Greenhouse gases also include the chlorofluorocarbon (CFC) family, several synthetic chemicals which have been implicated in the destruction of the Earth's stratospheric ozone layer. The CFCs trap heat in the atmosphere even more efficiently than methane, contributing about 20 percent as much greenhouse forcing as atmospheric carbon dioxide. Molecule for molecule, CFCs trap thousands of times as much heat as carbon dioxide. A molecule of sulfur hexafluoride, one of the chemicals subject to controls under the Kyoto Protocol, is 23,900 times more potent over 100 years than a molecule of carbon dioxide. Most CFC use is being phased out under international protocols, but its effects "will only decrease very slowly next century," according to John T. Houghton (1997, 37).

Levels of energy use and efficiency vary widely around the world. In 1990, the United States generated more carbon per capita (5.03 tons per person in 1987) than any other nation, with Canada (4.24), Australia (4.00), the Soviet Union (3.68), and Saudi Arabia (3.60) also above the world's average, which was 1.08. Zaire (0.03), Nigeria (0.09), Indonesia (0.16), and India (0.19) generated the least. By the late 1980s, the United States was producing six times as much economic activity per unit of carbon dioxide generated as China. The United States, in turn, used three times as much energy per unit of output as France and Japan (National Academy of Sciences 1991, 8). With regard to energy efficiency, the world average in 1987 was 327 grams of carbon produced per dollar of gross national product (GNP). The United States produced a dollar of GNP with 276 grams of carbon emissions, while worldwide, this measure ranged from 147 in Italy and 156 in Japan to 655 in India and 2,024 in China. The figures for China and India were inflated by widespread use of dirty, low-power coal. This type of usage is accelerating and will play a major role in worldwide greenhouse gas levels during the twenty-first century (Whalley and Wigle 1990, 238).

A baby born into the automated culture of the United States of America in the year 2000 consumed about a hundred times the natural resources (including fossil fuel energy) as a baby born in Bangladesh. The wealthiest one-quarter of the world's population consumed 80 percent of its aggregate energy resources. The average resident of an industrialized country consumed the energy contained in roughly thirty-two barrels of oil per year, about seven times the average consumption by residents of third world countries (Silver and DeFries 1990, 53). According to Anita Gordon and David Suzuki, "The average North American uses the energy equivalent of 10 tons of coal a year. Bangladeshis use less than 100 kilograms (220 pounds)" (1991, 106). In other words, the average North American (in the United States and Canada) contributes as much greenhouse effluent to the atmosphere as sixty Angolans or twenty-five residents of India.

The type of coal-powered industrialization that began in Britain three centuries ago is now taking place in China and India. The major ecological difference is that about 20 million people inhabited England during the dawn of its industrial revolution. China's population at the dawn of the twenty-first century was about 1.3 billion, and India's was nearing 1 billion.

China surpassed the United States as the world's largest burner of coal during the late 1980s; by 1990, China was responsible for 10 percent of the world's carbon dioxide emissions. By the middle of the twenty-first century, China might become the world's largest national source of greenhouse gases. Much of this emission load will be produced by the development of energy systems in urban areas. In the late 1980s, according to Lu Ying-zhong of Beijing's Institute for Techno-economics and Energy Systems, only 10 percent of China's oil and coal was being consumed in rural areas, where 80 percent of the people live (Oppenheimer and Boyle 1990, 141). World levels of greenhouse gases might jump as rural China hooks into power grids fueled by its extensive deposits of low-energy coal.

HISTORY OF THE GREENHOUSE EFFECT AS AN IDEA

The fossil-fueled industrial revolution was born in England. As coal-fired industry (as well as home heating and cooking) filled English skies with acrid smoke, some English homeowners protested coal's use as a fuel. Others described the horrors of coal mines, into which children as young as six were sent to work. Queen Elizabeth sometimes forbade the burning of coal in London while Parliament was in session. In 1661, John Evelyn wrote a book that complained about the noxious nature of coal smoke.

The coal-burning steam engine was invented by Thomas Newcomen in 1712, and it was refined into a form that was widely adaptable for industrial processes by James Watt, beginning in 1769. Within a century of industrialism's first stirrings, during the 1820s, Jean Baptiste Joseph Fourier, a Frenchman, compared the atmosphere to a greenhouse. During the 1860s, John Tyndall, an Irishman, developed the idea of an "atmospheric envelope," suggesting that water vapor and carbon dioxide in the atmosphere are responsible for retaining heat radiated from the sun. Tyndall also wrote that climate might warm or cool based on the amount of carbon dioxide and other gases in the atmosphere. Tyndall, who speculated in 1861 that a fall in carbon dioxide levels could have accounted for the ice ages, was the first person to make quantitative, spectroscopic measurements showing that water vapor and carbon dioxide absorb thermal radiation and could therefore trap solar heat in the atmosphere.

In 1896, Savante August Arrhenius, a Swedish chemist, published a paper in *The London, Edinburgh, and Dublin Philosophical Magazine and Journal of Science* titled "On the Influence of Carbonic Acid in the Air upon the Temperature of the Ground." In his paper, Arrhenius theorized that a rise in the atmospheric level of carbon dioxide could raise the temperature of the air. He was not the only person thinking along these lines at the time; Swedish geologist Arvid Hogbom had delivered a lecture on the same idea three years earlier, which Arrhenius incorporated into his article. Arrhenius was a well-known scientist in his own time, not for his theories describing the greenhouse effect but for his work on electrical conductivity, for which he was awarded a Nobel Prize in 1903. Later in his life, Arrhenius directed the Nobel Institute in Stockholm. His work in global-warming theory was not much discussed during his own life, however. Arrhenius, using the available measurements of absorption and transmission by water vapor and carbon dioxide, developed the first quantitative mathematical model of the Earth's greenhouse effect and obtained results of acceptable accuracy by today's standards for equilibrium climate sensitivity to carbon dioxide changes.

Arrhenius developed his theory through the use of equations by which he calculated that a doubling of carbon dioxide in the atmosphere would raise air temperatures about 10 degrees F. Arrhenius thought 3,000 years would have to pass before human-generated carbon dioxide levels would double. He applauded the possibility of global warming, telling audiences that a warmer world "would allow all our descendants, even if they only be those of a distant future, to live under a warmer sky and in a less harsh environment than we were granted" (Christianson 1999, 115). In his book *Worlds in the Making* (1908) Arrhenius wrote: "By the influence of the increasing percentage of carbonic acid in the atmosphere, we may hope to enjoy ages with more equable and better climates, especially as regards the colder regions of the Earth, ages when the Earth will bring forth much more abundant crops than at present for the benefit of rapidly propagating mankind" (p. 115).

Arrhenius' ideas were not widely discussed, but they did not completely die during the early twentieth century. Alfred J. Lotka, an American physicist, warned in 1924, "Economically we are living on our capital; biologically, we are changing radically the complexion of our share in the carbon cycle by throwing into the atmosphere, from coal fires and metallurgical furnaces, ten times as much carbon dioxide as in the process of breathing" (Oppenheimer and Boyle 1990, 35). Calculating on the basis of fossil fuel use in 1920, at the beginning of the automotive age, Lotka ventured an estimate that the level of carbon dioxide in the atmosphere would double in 500 years because of human activities, one-sixth of the time period earlier forecast by Arrhenius.

By the late 1930s, the prospect of global warming was catching the eye of G. D. Callendar, a British meteorologist, who gathered records from more than 200 weather stations around the world to propose that the Earth had warmed 0.4 degrees C between the 1880s and the 1930s because of carbon dioxide emissions by industry (Callendar 1938). While Callendar's assertions were met with skepticism by many English scientists at the time, he was laying the foundation for modern-day efforts to make more precise measurements of atmospheric trace-gas trends and radiative properties and to design more capable climate models to simulate climate change.

Two decades after Callendar's efforts, in 1956, Gilbert Plass, a scientist at Johns Hopkins University in Baltimore, suggested that carbon dioxide is an important influence on climate. He also projected that burning of fossil fuels would raise the global temperature 1.1 degrees C (2.0 degrees F) by the end of the century, very close to the actual worldwide increase.

During 1957, Roger Revelle and Hans Suess warned, as part of the International Geophysical Year, "Human beings are now carrying out a large-scale geophysical experiment of a kind that could not have

Dr. Charles D. Keeling. Courtesy of the Scripps Institution of Oceanography.

happened in the past nor be reproduced in the future. Within a few centuries we are returning to the atmosphere and oceans the concentrated organic carbon stored in sedimentary rocks over hundreds of millions of years" (Christianson 1999, 155–156). Revelle would become known to the world years later as the mentor of a graduate student, Albert Gore, who, in 1992 (a year after Revelle died), was elected vice president of the United States. The same year, Gore published a book, *Earth in the Balance*, which argued for mitigation of the greenhouse effect. Until the late 1950s, scientists had no reliable records of carbon dioxide and other greenhouse gas levels in the atmosphere.

At about the same time that Revelle and Suess issued their warning, Charles David Keeling began to assemble documentation that indicated that the worldwide level of carbon dioxide had risen to about 315 ppm, compared to about 280 ppm at the end of the previous century. Keeling's calculations also indicated that the level was continuing to rise steadily. When Keeling decided to measure the concentration of carbon dioxide in the atmosphere, he first had to construct a machine to obtain readings in parts per million. No such machine existed at the time. Keeling worked on his "manometer" for a year, building the machine from an old blueprint at the California Institute of Technology in Pasadena. Keeling's first readings, on the roof of a Caltech laboratory, showed an atmospheric concentration of 310 ppm. Keeling next took his manometer on family vacations, recording readings. What he found contradicted scientific assumptions of the time, which held that carbon dioxide levels would vary widely, depending on local sources of the gas. Instead, he found that carbon dioxide readings were very similar in different places, indicating its rapid diffusion throughout the atmosphere.

"I decided all the data in the literature was wrong," Keeling later told William K. Stevens, a science reporter for the *New York Times* (Stevens 1999, 140). After Keeling and his associates had taken a number of readings for more than a year, Keeling made a discovery that confirmed Revelle and Suess' assertions. The readings on his manometer were rising steadily, year by year. The resulting graph, which plots the level of carbon dioxide in the atmosphere, came to be known among scientists as the "Keeling curve."

Keeling also was discovering that atmospheric carbon dioxide levels vary annually, with lower readings in spring and summer (when plants are respiring oxygen on the large land masses of the Northern Hemisphere), and higher levels during fall and winter, when many plants are dormant and more carbon is being released as a result of vegetative decay. This annual cycle can vary by as much as 3 percent in the Northern Hemisphere, compared to 1 percent in the Southern Hemisphere, where the dominance of oceans mutes seasonal cycles and restricts variability in the carbon dioxide level.

Until the work of Revelle, Suess, and Keeling, most scientists who studied carbon dioxide levels in the atmosphere assumed that the oceans absorbed all of the extra carbon dioxide that human activities were injecting into the air. The readings of Keeling and his associates showed that human activity was steadily raising the level of carbon dioxide in the atmosphere more quickly than the oceans or other "sinks" of the gas could absorb it.

By the early 1960s, a growing number of scientists were watching Keeling's carbon dioxide readings from the Mauna Loa observatory. During 1963, the Conservation Foundation issued a warning in a report titled "Implications of Rising Carbon Dioxide Content of the Atmosphere," which concluded: "It is estimated that a doubling of carbon-dioxide content in the atmosphere would produce a temperature rise of 3.8 degrees [C]" (Kellogg 1990, 99). Revelle played a key role in the creation, during 1970, of the Scientific Committee on Problems of the Environment as part of the International Council of Scientific Unions (ICSU). He also suggested the objective for the ICSU's International Geosphere-Biosphere Program in 1986: "To describe and understand the interaction of the great global physical, chemical, and biological systems regulating planet Earth's favorable environment for life, and the influence of human activity on that environment" (Malone, Goldberg, and Munk n.d.).

GLOBAL WARMING BECOMES A POLITICAL ISSUE

The modern debate over whether the lower atmosphere is warming as a result of human activity began in policy circles during 1979, after a small number of well-known scientists reported to the Council on Environmental Quality that "man is setting in motion a series of events that seem certain to cause a significant warming of world climates unless mitigating steps are taken immediately" (Pomerance 1989, 260). At the same time, the National Academy of Sciences initiated a study of the greenhouse effect. Also during 1979, in the United States, the President's Council on Environmental Quality mentioned global warming: "The possibility of global climate change induced by an increase of carbon dioxide in the atmosphere is the subject of intense discussion and controversy among scientists" (Anderson 1999).

An alarm regarding global warming was sounded in the May 3, 1979, issue of the British journal *Nature*: "The release of carbon dioxide to the atmosphere by the burning of fossil fuels is, conceivably, the most important environmental issue in the world today" (Bernard 1993, 6). The term "global warming" was first defined and used in a scientific sense by James Hansen and colleagues in *Science*, in a paper titled "Climate Impact of Increasing Atmospheric Carbon Dioxide" (Hansen et al. 1981, 957–966). "President Ronald Reagan's Department of Energy responded by canceling funding to the Goddard Institute for Space Studies, which Hansen directs, because they didn't like the results that we were getting" (Lacis 2005).

At about the same time, a study conducted by a scientific team chaired by meteorologist Jule Charney estimated that doubling the carbon dioxide level in the atmosphere would raise the average global temperatures by about 3 degrees C, plus or minus 1.5 degrees C. In 1983, the United States Environmental Protection Agency released a report, "Can We Delay a Greenhouse Warming." A National Academy of Sciences report, issued in the same year, stated, "We do not believe that the evidence at hand about CO_2-induced climate change would support steps to change current fuel-use patterns away from fossil fuels" (Pomerance 1989, 261).

Until the early 1980s, those who argued that human contributions to the greenhouse effect had raised (or would raise) the temperature of the atmosphere near the Earth's surface had a statistical problem. Starting in 1940 and lasting until about 1975, yearly average global temperatures actually fell slightly. After 1975, temperatures began a steady rise. During 1985, Veerabhaadran Ramanathan, Ralph Cicerone, and their colleagues at the National Center for Atmospheric Research proposed that trace gases other than carbon dioxide could be as dangerous vis-à-vis greenhouse warming as carbon dioxide. Also, by 1985, Claude Lorius was beginning to demonstrate that lower carbon dioxide levels in the atmosphere strongly correlated with lower temperatures during the last Ice Age.

The potential impact of warming as a result of infrared forcing was raised again at United Nations–sponsored conferences in Villach, Austria, during the middle 1980s. Shortly after these conferences, senators David Durenberger of Minnesota and Al Gore of Tennessee called for an international "Year of the Greenhouse" to raise the issue in public consciousness. Gore already had played a role in congressional hearings on the issue in 1982 and 1984, when he was serving in the House of Representatives.

As a political issue in the United States of America, global warming came of age during the notably hot summer of 1988. That year provided something of a wake-up call in the debate over global warming because it was the warmest since reliable records had been kept in the middle of the nineteenth century. During 1988, 400 electrical transformers in Los Angeles blew out on a single day as temperatures rose to 110 degrees F. Two thousand daily temperature records were set that year in the United States. Widespread heat and drought caused some crop yields in the Midwest to fall between 30 and 40 percent. In Moscow, Russians escaping their hottest summer in a century flocked to rivers and lakes, where they drowned in record numbers (Christianson 1999, 197).

During 1988, Colorado Senator Timothy E. Wirth, whose hearings on global warming the previous winter had drawn little attention, played the weather card. He called another hearing, this time during the summer. As it happened, the hearing convened on a particularly hot, humid day in Washington, D.C., during which the temperature reached a record 101 degrees F. At Wirth's hearing, James Hansen, head of the federal government's Goddard Institute for Space Studies, testified that the unusually warm temperatures of the 1980s were an early portent of global warming, which was caused by the burning of fossil fuels and not solely a result of natural variation. Hansen's remarks became front-page news nationwide within hours. Hansen also continued a running battle, during the Reagan and Bush administrations, to raise the political salience of global warming despite funding cuts (and threats of cuts) of the Goddard Institute. The Office of Management and Budget forced Hansen to censor the severity of his findings several times. The pressure was so intense that Hansen sometimes asked to testify as a private citizen rather than as a federal employee (Hansen 1989).

In the meantime, the scientific debate over global warming was intensifying. By the end of 1988, the U.N. General Assembly had approved the creation of the Intergovernmental Panel on Climate Change. A year later, Hansen said that it was "time to cry wolf":

> When is the proper time to cry wolf? Must we wait until the prey, in this case the world's environment, is mangled by the wolf's grip? The danger of crying too soon, which much of the scientific community fears, is that a few cool years may discredit the whole issue. But I believe that decision-makers and the man-in-the-street can be educated about natural climate variability....A greater danger is to wait too long. The climate system has great inertia, so as yet we have realized only a part of the climate change which will be caused by gases we have already added to the atmosphere. Add to this the inertia of the world's energy, economic, and political systems, which will affect any plans to reduce greenhouse gas emissions. Although I am optimistic that we can still avoid the worst-case climate scenarios, the time to cry wolf is here. (Nance 1991, 267–268)

Hansen elaborated: "I said three things [in 1988]. The first was that I believed the Earth was getting warmer and I could say that with 99 percent confidence. The second was that with a high degree of confidence we could associate the warming and the greenhouse effect. The third was that, in our climate model, by the late 1980s and early 1990s, there's already a noticeable increase in the frequency of drought" (Parsons 1995, 7).

Between June 27 and 30, 1988, as the Earth's warmest summer on record (to that time) was getting under way, more than 300 leaders in science, politics, law, and environmental studies gathered in Toronto at the invitation of Canada's government to address problems related to climate change, including prospects of global warming. A scientific consensus was forming around the idea that human activity already was altering the Earth's atmosphere at an unprecedented rate. A consensus statement issued by the Toronto climate conference asserted, "There can be a time lag of the order of decades between the emission of gases into the atmosphere and their full manifestation in atmospheric and biological consequences. Past emissions have already committed planet Earth to a significant warming" (Ferguson 1989, 48).

During the 1992 presidential campaign in the United States, candidate Bill Clinton criticized the George H. W. Bush administration's refusal to join in worldwide diplomatic efforts to reduce emissions of greenhouse gases. Once elected, Clinton's first budget proposed a carbon tax, a measure that was quickly dropped under pressure from Republicans in Congress, where the tax died in committee. Meanwhile, a climate convention signed by 161 countries at the Conference on Environment and Development in Rio de Janiero, during 1992, contained a directive, in Article 2, favoring stabilization of greenhouse gases "at levels and on a time scale that do not produce unacceptable damage to ecosystems and that allow for sustainable economic development" (Woodwell and MacKenzie 1995, v).

At about the same time, several studies (examples being Easterling et al. [1997] and Karl et al. [1993]) indicated that daily minimum temperatures had increased more rapidly during much of the twentieth century than daily maximums. Easterling reported that between 1950 and the mid-1990s, daily minimums increased at a rate of 1.86 degrees C per century, while maximums increased 0.88 degrees C (less than half as much) during the same period. A study by Henry F. Diaz and Raymond S. Bradley supported climate-model forecasts that daily minimum temperatures will rise more rapidly than maximums in a warmer world.

Diaz and Bradley, who studied temperature changes during the twentieth century at high-elevation sites, wrote, "The signal appears to be more closely related to increases in daily minimum temperatures than changes in the daily maximum. The changes in surface temperature vary spatially, with Europe (particularly western Europe) and parts of Asia displaying the strongest high-altitude warming during the period of record" (Diaz and Bradley 1997, 253).

A number of legislative bodies in different parts of the world took initiatives to limit greenhouse gas emissions soon after the memorably hot summer of 1988. During 1989, the Netherlands passed a National Environmental Policy Plan, which required a freeze on carbon dioxide emissions at 1989 and 1990 levels. The parliament of Norway decided to limit carbon dioxide emissions in that country to 1989 levels by the year 2000, with a decline in emissions mandated after that. The Vermont State Legislature enacted a law outlawing automobile air conditioners after 1993 unless a substitute could be developed to replace CFCs. On June 13, 1990, just before Germany was reunified, the West German Cabinet committed the country to a 25 percent reduction in greenhouse gases, based on 1987 levels, by the year 2005. Later, the unified government stood behind these limits but added allowances for energy-inefficient industries in what had been East Germany.

Stephen H. Schneider's *Global Warming: Are We Entering the Greenhouse Century?* (1989) was one of the first popular treatments of global warming in book form. During the next several years, Schneider became a leading scientific voice in public debates over the issue. By the late 1990s, Schneider was a professor in the Department of Biological Sciences and a senior fellow at the Institute for International Studies at Stanford University. He was honored in 1992 with a MacArthur Fellowship for his ability to integrate and interpret the results of global climate research through public lectures, seminars, classroom teaching, environmental assessment committees, media appearances, congressional testimony, and research collaboration with colleagues. Schneider served as a consultant to several federal agencies as well as a member of the White House staff in the Nixon, Carter, Reagan, Bush, and Clinton administrations. In 1975, Schneider founded the interdisciplinary journal

Climatic Change and served as its editor. Schneider also edited *The Encyclopedia of Climate and Life* (1996) and *Laboratory Earth: The Planetary Gamble We Can't Afford to Lose* (1997).

Schneider's *Global Warming: Are We Entering the Greenhouse Century?* ends with a passage that sounded rather prescient at the year 2004:

"Are we now entering the Greenhouse Century?" I asked in the subtitle of this book. It should be clear by now that I believe we've been in it for a while already, but admit that it will take a decade or so more of record heat, forest fires, intense hurricanes, or droughts to convince the substantial number of skeptics that still abound. Unfortunately, while the antagonists debate, the greenhouse gases keep building up in the atmosphere. I wonder what we will say to our children when they eventually ask what we did—or didn't do—to create the Greenhouse Century they will inherit. (Schneider 1989, 285)

Schneider continued: "I strongly suspect that by the year 2000 increasing numbers of people will point to the 1980s as the time the global warming signal emerged from the natural background of climatic noise" (Schneider 1989, 32). When he was asked whether human activities had assumed a dominant role in climate change, Schneider said: "I'm not 99 percent sure, but I am 90 percent sure. Why do we need 99 percent certainty when nothing else is that certain? If there were only a 5 percent chance the chef slipped some poison in your dessert, would you eat it?" (Landsea 1999). Schneider argued that everyone would be 100 percent certain that scientists' beliefs about climate change are accurate only when they observe dramatic, adverse changes in climate. By then, he asserted, it will be too late to take remedial measures.

PRESENT EMISSIONS = FUTURE WARMING

A major concern of Schneider and other scientists is that present-day political debates respond mainly to the degree of warming that people feel in the present tense. In reality, however, the warmth felt today reflects fossil fuel emissions of several decades ago. The climate system is never actually in thermodynamic equilibrium. Rather, it is forever playing catch-up with the daily and seasonal variations of incoming sunlight, as the ground tries to come into thermal equilibrium with changes in solar radiation. This is where the heat capacity of the ocean, ground, and atmosphere come into play, as well as the thermal opacity of the atmosphere, which regulates how readily heat energy from the ground can be radiated to space. The time it takes the system to get to a new equilibrium is expressed in terms of a time constant, or "e-folding" time—that is, the length of time that it takes the system to reach approximately 63 percent of its final equilibrium temperature. Mathematically, it takes forever to reach "true" equilibrium, but practically, after a few e-folding times, the system can be said to be in effective equilibrium. The e-folding time of the atmosphere is a few months, the mixed layer of the ocean a few years, and the total (deep) ocean a few hundred years. Present climate has accumulated about 0.5 watts per meter squared of unrealized warming.

In 1989, Ramanathan stated:

The rate of decadal increase of the total radiative heating of the planet is now about five times greater than the mean rate from the early part of this century. Non-CO_2 trace gases in the atmosphere are now adding to the greenhouse effect by an amount comparable to the effect of CO_2 increase. . . . The cumulative increase in the greenhouse forcing until 1985 has committed the planet to an equilibrium warming of about 1 to 2.5 degrees C. (Ramanathan 1989, 241)

Ramanathan continued:

The climate system cannot restore the equilibrium instantaneously, and hence the surface warming and other changes will lag behind the trace-gas increase. Current models indicate that this lag will range [from] several decades to a century. However, analyses of temperature records of the last 100 years as well as proxy records [of] paleoclimate changes indicate that climate changes can also occur abruptly instead of a gradual return to equilibrium as estimated by models. The timing of the warming is one of the most uncertain aspects of the theory. (Ramanathan 1989, 245)

Peter Ciborowski projected, "Within 50 years, we will be committed to a mean global temperature rise of 1.5 degrees C to 5 degrees C. And if no attempt is made to slow the rate of increase, we could be committed to another 1.5 degrees C to 5 degrees C in another 40 years" (Ciborowski 1989, 227–228).

2 SCIENTIFIC RESEARCH: THE ISSUE'S COMPLEXITY

INTRODUCTION

Global warming research has become big-ticket science. Nearly every week, the most recent developments are carried in major scientific journals such as *Science* and *Nature* as well as publications for specialists in several fields, including geophysics, meteorology, and several other earth sciences. While the public debate often carries echoes of nearly doctrinal certainty from nonscientists, controversies within disciplines often flare with uncertainty. Research has called into question some of the most fundamental assumptions of climate diplomacy. For example, the Kyoto Protocol relies on assumptions that planting forests will curb greenhouse gases. Research indicates that this may not be the case. Lush grassland may be a better carbon "sink" (absorber) under some circumstances.

The basics of the carbon cycle are not fully understood. Until recently, scientists have not been able to account for all the carbon used in the worldwide cycle. The relationship of a certain level of greenhouse gases in the atmosphere to a given amount of warming over an assumed length of time is open to pointed debate. Other factors (the Earth's orbit, solar insolation, land-use patterns, and others) are provided more or less weight in debates regarding how quickly temperatures might rise as consumption of fossil fuels raises greenhouse gas levels in the atmosphere. How is warming related to drought cycles? What role do clouds (including contrails, the exhaust of jet aircraft) play in the general warming of the Earth? The role of aerosols (the most important probably being black soot) is only beginning to be understood and factored into forecast models.

This chapter concludes with a consideration of the history of Venus, known today as one of the most hellish environments anywhere, an example of what can happen to a planet in the grip of a runaway greenhouse effect. A scientific case has been made that, not long ago by the standards of geologic time, Venus' atmosphere much more closely resembled Earth's. The geophysical career of Venus long has been a subject of inquiry among leading scientists of global warming, such as James E. Hansen, director of NASA's Goddard Institute for Space Studies, who began his professional life as a student of that planet. The climatic transect of Venus may be a cautionary tale for our Earth.

THE GLOBAL SCIENTIFIC CONSENSUS

Basic scientific concern revolves around the rate at which carbon dioxide and other greenhouse gases are accumulating in the atmosphere, which has been rising steadily, for the most part, since the dawn of the industrial age. When systematic measurements of the carbon dioxide level began during the late 1950s, the annual rise (which varies year to year) was about 1 part per million (ppm). The level, which before the industrial age peaked at about 280 ppm during interglacial periods, was about 315 ppm at that time. During the 1990s, with the level approaching 370 ppm, the rate of increase was about 1.8 ppm per year. In March 2004, the rate of increase (at 379 ppm) was measured at 3 ppm, compared to the previous year. Measurements are taken at a National Oceanic and Atmospheric Administration laboratory at the Mauna Loa Observatory, 11,141 feet above sea level on the island of Hawaii.

A detailed study of climatic history indicates that the Earth is warmer now than at any time in the past 2,000 years. Philip Jones of the University of East Anglia's Climate Research Unit coauthored a study with

Michael Mann of the University of Virginia that studied tree rings, ice cores, and lake sediments as far apart as Tibet, Greenland, and western North America, as well as European tree rings and fossil shells from America's Chesapeake Bay. The study, published in *Geophysical Research Letters* (Mann and Jones 2003) indicated "no sign of global warming slowing down and predicted the world is likely to be up to 3 degrees C. warmer within 100 years" (Ingham 2003, 19). The scientists said: "These reconstructions indicate late twentieth century warmth is unprecedented for at least roughly the past 2,000 years for the Northern hemisphere" (p. 19). The study suggests that in each century temperatures fluctuate an average of 0.2 degrees C, but the last twenty-five years have shown a "dramatic" acceleration in warming (p. 19).

WHERE HAS ALL THE CARBON GONE? MYSTERIES OF THE CARBON CYCLE

Scientists have been debating how much human-produced carbon dioxide is being absorbed by natural sinks and which of these are most important. On that score, considerable disagreement has become evident, with various specialists advancing their favored carbon sinks. What is known is that between 1800, which was roughly the beginning of the industrial age, and the present day, increased carbon dioxide in the atmosphere has amounted to only about half of what has been emitted.

According to some calculations (Sabine et al. 2004, 367), most of this "missing carbon" has been absorbed by the world's oceans, along a "pathway" by which at least some of the excess carbon ends up as calcium carbonate ($CaCO_3$) in the shells of marine creatures (Takahashi 2004, 352). While the oceans have absorbed only about one-third of their potential capacity for carbon dioxide, the rising proportion of this gas in the oceans has begun to change the chemistry of seawater, including its acidity, in ways that could eventually pose problems for marine life (Feely et al. 2004, 362).

Other scientists have asserted that the land is a major sink for the missing carbon. The United States produces more than 5 billion tons of carbon dioxide each year, and some scientists believe that mainland U.S. ecosystems soak up 10 to 30 percent of it, more than models say they should be absorbing. This proportion also has been increasing over time. Much of this sequestration might be taking place as a result of new plant growth, possibly regrowth of land previously logged or accelerated growth caused by global warming. One study has suggested that "increased rainfall and humidity documented in the continental United States might be the single most important factor spurring increased plant growth" (Lovett 2002, 1787).

According to Steven C. Wofsy, writing in *Science*, increasing amounts of anthropogenic (human-produced) carbon dioxide are being removed by forests and other components of the biosphere (Wofsy 2001, 2261). According to Wofsy, roughly 25 percent of anthropomorphic carbon dioxide is being absorbed by forests, which is roughly 2 million metric tons of carbon per year. Scientists' models use forest inventories that have trouble accounting for all of this stored carbon. Perhaps some models include only parts of forests that are commercially valuable, missing carbon sequestered by "woody debris, soil, wood products preserved in landfills, and woody plants that have encroached on grasslands because of the long-term suppression of natural fires," according to Wofsy (2001, 2261; Pacala et al. 2001, 2316). Wofsy cited estimates that up to 75 percent of the carbon sequestered in the United States "is found in organic matter that is not inventoried" (Wofsy, 2001, 2261). China also has experienced increases in carbon sequestration despite population pressure on its forests. Reforestation and afforestation (planting of trees in areas that did not previously have them) has been official policy there since the late 1970s, "motivated by the desire to restore degraded ecosystems for flood and erosion control" (Wofsy 2001, 2261; Fang et al. 2001, 2320).

Ramakrishna Nemani and colleagues at the University of Montana School of Forestry in Missoula investigated warming-related plant growth with a grant from NASA, using climate data from 1950 to 1993. They estimated that increases in rainfall were responsible for about two-thirds of plants' additional growth. Increased rainfall allows more water for growth, while higher humidity allows plants to grow more rapidly by opening their pores wider so they can take in more carbon dioxide, according to Steven Running of the University of Montana, one of the study's coauthors. The authors of the study found that increases in plant growth correlate with increases in rainfall across the United States (Lovett 2002, 1787). The study suggests that planting new forests to increase the effects of carbon sinks are, by themselves, "overly naïve" (p. 1787).

Nemani and colleagues found that "global changes in climate have eased several critical climatic constraints to plant growth, such that net primary production increased 6 percent (3.4 petagrams of carbon over 18 years) globally. The largest increase was in tropical ecosystems. Amazon rainforests accounted for 42 percent of the global increase in net primary production, owing mainly to decreased cloud cover and the resulting

Amazon rainforest. Courtesy of Getty Images/PhotoDisc.

increase in solar radiation" (Nemani et al. 2003, 1560). Although increases were most marked in tropical ecosystems, "significant growth stimulation [was recorded] in both the tropics and the northern high-latitude ecosystems" (p. 1562).

Studies surveyed by D. S. Schimel and colleagues in *Nature* indicated that the terrestrial biosphere was "neutral with respect to net carbon exchange" during the 1980s but became a net carbon sink during the 1990s. This change is believed to have been related to regrowth of forests on previously abandoned agricultural land in temperate-zone areas, fire prevention, and longer growing seasons resulting from rising temperatures (Schimel et al. 2001, 169). After surveying the field, Schimel and colleagues acknowledged considerable uncertainties "as to the magnitude of the sink in different regions and the contribution of different processes" (p. 169).

Some climate models anticipate that, until about 2050, forests, especially in the Northern Hemisphere, will absorb a declining amount of the carbon dioxide that humans pump into the air. After nearly eight years of experiments, however, Duke University researchers are less optimistic. "Our results are clear. Forests will be a modest, if not disappointing, sink for carbon dioxide," said William Schlesinger, a Duke biogeochemist who led one of two teams reporting results in *Nature* (Spotts 2001). The researchers used an array of towers to spray three stands of loblolly pine with enhanced carbon dioxide, then compared the pines with "control stands" of trees that were exposed to only the carbon dioxide of ambient air. "The trees initially responded to higher CO_2 levels by producing 24 to 34 percent more wood than the control trees, but growth tapered off to marginal increases after about three years," they found (Spotts 2001).

Writing in *Nature*, Paul A. del Giorgio and Carlos M. Duarte commented that the role of the oceans as well as forests as major carbon sinks has come into question:

A key question when trying to understand the global carbon cycle is whether the oceans are net sources or sinks of carbon. This will depend on the production of organic matter relative to the decomposition due to biological respiration. Estimates of respiration are available for the top layers, the mesopelagic layer, and the abyssal waters and sediments of various ocean regions. Although the total open ocean respiration is uncertain, it is probably substantially greater than most current estimates of particulate organic matter production. Nevertheless, whether the biota act as a net source or sink of carbon remains an open question. (del Giorgio and Duarte 2002, 379)

HOW DOES CARBON DIOXIDE RELATE TO TEMPERATURE?

A key question—some argue it is *the* key issue—in climate change is determination of how the level of carbon dioxide and other greenhouse gases in the atmosphere relate to a specific rise or fall in temperature after other "forcings" have been factored out. For example, how much will global temperatures rise, on average, if the carbon dioxide level doubles, all other factors being equal (which they rarely are)? Humankind has been altering the Earth's climate for several thousand years; cultivation of rice, for example, raises methane levels, and biomass burning increases the amount of carbon dioxide in the air (White 2004, 1610). In effect, human interference with the carbon cycle began, in a small way, the first time that a human being cooked his dinner over an open fire.

David W. Lea, writing in *Journal of Climate*, said that several models have produced varying answers to this question. To sharpen the debate, Lea turned to paleoclimatic data describing both tropical sea-surface temperatures and Antarctic ice cores. Using proxy records from the eastern equatorial Pacific Ocean and Vostok (Antarctic) ice cores, Lea arrived at an estimate of "a tropical climate sensitivity of 4.4 to 5.6 degrees C (error estimated at plus or minus 1.0 degree C) for a doubling of atmospheric CO_2 concentration" (Lea 2004, 2170). This result, according to Lea, "suggests that the equilibrium response of tropical climate to atmospheric CO_2 changes is likely to be similar to the upper end of available global predictions from global models" (p. 2170).

STORIES TOLD BY ICE CORES

A 3.19-kilometer column of ice drilled from eastern Antarctica by European scientists with the European Project for Ice Coring in Antarctica (EPICA), a consortium of laboratories and Antarctic logistics operators from ten nations during 2004 provided researchers with the oldest and most detailed record of climate change ever obtained, stretching back 740,000 years. The ice core indicated that today's greenhouse gas concentrations in the atmosphere are by far the highest for at least 440,000 years. This figure may become larger as scientists examine older parts of the new ice core. Smaller increases in greenhouse gas levels that were provoked by natural causes have been followed by significant rises in global temperatures. Ice cores are valuable records of Earth's past climate because they record variations in temperatures as well as concentrations of gases such as carbon dioxide and methane that contribute to the greenhouse effect. The new core doubled the available paleoclimatic record when it was completed during 2005.

By late 2005, scientists were reporting analyses of carbon dioxide, methane, and nitrous oxide in Antarctic ice cores indicating that present-day levels of the gases in the atmosphere are higher than at any point in at least the last 650,000 years (Siegenthaler et al. 2005, 1313; Spahni et al. 2005, 1317). These results extended the previous record from 440,000 years before the present. Today's carbon dioxide level, at 380 ppm and still rising, is 27 percent higher than at any point in the ice-core record, according to Thomas Stocker of Switzerland's University of Bern, one of several scientists comprising the European Project for Ice Coring in Antarctica, which conducted the most recent study ("Study of Ancient" 2005, 4-A). Levels of methane are similarly, unusually high. According to Geoscientist Edward Brook of Oregon State University, "There's no natural condition that we know of in a really long time where the greenhouse-gas levels were anything near what they are now" ("Study of Ancient" 2005, 4-A; Brook 2005, 1285).

Furthermore, the rate at which humankind's burning of fossil fuels is changing the composition of the atmosphere is extraordinary by natural standards. "The rate of increase [in greenhouse gases] is more than 100 times faster than any rate we can detect from the ice cores we have seen so far," said Thomas Stocker, a member of EPICA (Henderson 2004, 4).

The new core confirms evidence from ocean sediments that the Earth has endured several ice ages during the last 740,000 years, each separated by a warmer interglacial period. While ice ages typically last about 100,000 years, the interglacials usually are much shorter, averaging (very roughly) 10,000 years each (Henderson 2004, 4). EPICA's findings indicate an "extremely strong" 100,000-year cycle for ice ages during at least the last 500,000 years, with the present interglacial most resembling another about 430,000 years ago (Wolff et al. 2004, 623). That interglacial lasted about 28,000 years, longer than most. Why has the present interglacial been longer than most? It may partially be a matter of "how much, where, and during what season the Sun's energy reaches the planet," according to one measurement the EPICA scientists are using (White 2004, 1610). The shape of the Earth's orbit around the sun (which determines Earth's distance from the Sun during different seasons) and its interaction with the seasonal cycle seem to fit the waxing and waning of ice ages noted in ice cores.

Because the present interglacial is about 12,000 years old, the EPICA scientists put very little credibility in the assertions of global warming skeptics that burning fossil fuels might forestall a new ice age any time soon. "Given the similarities between this earlier warm period and today, our results may imply that without human intervention, a climate similar to the present one would extend well into the future" (Wolff et al. 2004, 623). Given rising levels of greenhouse gases, however, the team finds any assumption that climate will remain stable "highly unlikely" (p. 627).

By June 2004, bubbles of air in the new ice core had been examined for greenhouse gas concentrations for its first 440,000 years. These samples indicated levels of carbon dioxide hovering between 200 ppm during ice ages and 280 ppm during interglacials. These small rises have been associated with the rising temperatures of the warmer periods. Since the industrial revolution, however, carbon dioxide concentrations have risen to about 380 ppm. The fact that human activity has raised the atmosphere's carbon dioxide level to about 30 percent above the natural range for the last 440,000 years or more provokes scientists to wonder just how much Earth's temperature will rise in coming years, especially if carbon dioxide levels continue to rise. "In everything we have got up to now, temperature and greenhouse gases are absolutely in step with each other," Wolff said. "I don't see any particular reason this shouldn't continue into the future. It is worrying" (Henderson 2004, 4).

SOIL WARMING MIGHT STIMULATE CARBON STORAGE

Some studies indicate that midlatitude forest soils lose small amounts of carbon in response to soil warming. These studies further indicate that soil conditions created by warming may actually enhance carbon storage by plants. Specifically, a report in *Ascribe Newsletter* suggested that stimulated carbon storage at the ecosystem level could slow the rate of climate change ("Study Shows Soil Warming" 2002). This report challenges assumptions in some climate models that project large, long-term releases of the greenhouse gas carbon dioxide in response to the warming of forest ecosystems.

This experiment, based in central Massachusetts, was led by Jerry M. Melillo, codirector of the Marine Biological Laboratory's Ecosystems Center. Melillo and colleagues found that while warming accelerated the decay of organic material and consequently the release of carbon dioxide from the soil, the response was limited and short-lived. Over the ten–year study period, only 11 percent of available soil carbon was lost to the atmosphere. The rest of the carbon was in a chemical form difficult for soil bacteria to break down and release. This phenomenon creates a built-in biochemical brake on the release of soil carbon, said Melillo ("Study Shows Soil Warming" 2002; Melillo et al. 2002, 2173).

"Here we show," wrote Melillo and colleagues, "that whereas soil warming accelerates soil organic matter decay and carbon-dioxide fluxes to the atmosphere, this response is small and short-lived for a midlatitude forest, because of the limited size of the labile soil carbon pool. We also show that warming increases the availability of mineral nitrogen to plants. Because plant growth in many midlatitude forests is nitrogen-limited, warming has the potential to indirectly stimulate enough carbon storage in plants to at least compensate for the carbon losses from soils. Our results challenge assumptions made in some climate models that lead to projections of long-term releases of soil carbon in response to warming of forest ecosystems" (Melillo et al. 2002, 2173).

This study also indicates that soil warming tends to increase nitrogen availability for plants. Because nitrogen is the limiting nutrient in many midlatitude forests, an increase in nitrogen availability, the researchers contend, creates the conditions required for plant growth, stimulating carbon storage in trees. They concluded that the amount of carbon stored in trees at their research site over the ten–year study period is at least equal to the carbon lost to the atmosphere from the soil ("Study Shows Soil Warming" 2002).

This study could be important to global climate modelers, according to Melillo, in part because some climate models show a runaway warming feedback based on an assumption that global warming will lead to increased decomposition in forest soils that will inject increased amounts of carbon dioxide into the atmosphere, contributing further to warming. "Our study shows that this may not be the way things would occur," Melillo said. "When creating models, it's important to build in positive and negative feedbacks and where they are occurring on earth in order to generate more realistic climate prediction. Modelers need to consider the mosaic nature of the land surface" ("Study Shows Soil Warming" 2002).

CAN TEMPERATURE FALL WHEN CARBON DIOXIDE LEVELS RISE?

During most of the Earth's history, temperatures have risen and fallen roughly in tandem with carbon dioxide levels. To add another mystery to the equation, however, this has not *always* been the case. Jan Veizer, Yves Godderis, and Louis M. Francois (2000, 698–701) presented a temperature record using the oxygen isotopic composition of tropical marine fossils describing periods in which there were mismatches between carbon dioxide levels and temperatures: the late Ordovician (440 million years ago) and the Jurassic and early Cretaceous (120 to 220 million years ago). Veizer and colleagues' findings indicated that, very occasionally, rising levels of carbon dioxide were accompanied by a *drop* in average temperatures. During the Jurassic, 145 million to 208 million years ago, for example, carbon dioxide levels might have been as much as ten times present levels, but temperature averages were cooler. Veizer believes that the sun might be a major driver of climate change, disrupting the relationship of temperature and CO_2 (p. 698). This conclusion is far from unquestioned, however. Some scientists question the climate proxies used in these studies, asserting that they might be producing inaccurate data (Kump 2000, 651; Pearson et al. 2001, 481).

Lee R. Kump has commented in *Nature*: "Either the CO_2 proxies are flawed, or our understanding of the relationship between CO_2 and climate . . . needs rethinking" (Kump 2000, 651). Kump believes that "lack of close correspondence between climate change and proxy indicators of atmospheric CO_2 might force us to reevaluate the proxies, rather than disallow the notion that substantially increased atmospheric CO_2 will indeed lead to marked warming in the future" (p. 652).

Climate modelers have long been perplexed by an apparent discrepancy between "climate models with increased levels of carbon dioxide [which] predict that global warming causes heating in the tropics [compared to] . . . investigations of ancient climates based on paleodata [that] have generally indicated cool tropical temperatures during supposed greenhouse episodes" (Pearson et al. 2001, 481). During the late Cretaceous (about 970 million years ago) and Eocene (about 50 million years ago) periods, for example, the poles were believed to have been ice-free and tropical sea-surface temperatures have been estimated to have been between 15 degrees and 23 degrees C, based on oxygen-isotope paleothermometry of surface-dwelling planktonic foraminifera shells, "which provide proxy information on ocean temperatures" (Kump 2001, 470). This data indicates an ice-free Earth in which tropical sea-surface temperatures were cooler than today in most areas—a very unusual circumstance by present-day standards. Either the Earth was very different at that time, or the proxies being used to estimate temperatures are flawed, or both.

Writing in *Nature*, Paul N. Pearson and colleagues questioned the validity of most such data "on the grounds of poor preservation and diagenetic alteration" (2001, 481). They presented new data "from exceptionally well-preserved foraminifera shells extracted from impermeable clay-rich sediments," which indicated temperatures of at least 28 degrees to 32 degrees C (compared to 25 to 27 degrees C today) in the tropics during these periods. (Foraminifera shells are valuable in paleoclimatic research because they can be dated.) Such estimates would indicate warming in the tropics consistent with the rest of the Earth during periods of high atmospheric carbon dioxide. These new data indicate that "for those parts of the Late Cretaceous and Eocene epochs that we have sampled, tropical temperatures were at least as warm as today, and probably several degrees warmer [allowing] us to dispose of the 'cool tropic paradox' that has bedeviled the study of past warm climates" (p. 486).

This study also confirms the relationship between past episodes of elevated atmospheric carbon dioxide and global warming. These results fit well with existing knowledge that temperature-sensitive organisms, such as corals, often were displaced northward during times when the atmosphere's carbon dioxide levels were relatively high (Kump 2001, 471). Such episodes are being used as predictors of conditions in the future, when "fossil-fuel burning is likely to drive atmospheric CO_2 to perhaps six times the pre-industrial level" (p. 470).

ARE FORESTS OVERRATED AS A CARBON SINK?

Forests have long been assumed to be an antidote to global warming; trees, because of their size, have been thought to be the most voracious consumers of carbon dioxide among Earth's flora. The Kyoto Protocol is shot through with this assumption, which might not always be true, according to research completed after the protocol was negotiated. According to some scientific studies, lush grasslands under some

Grassy meadow in the Santa Ynez Mountains. Courtesy of Corbis.

conditions remove more carbon dioxide from the atmosphere than forests.

Previous estimates of the amount of carbon stored by trees and shrubs might have been too high. According to an account by Cat Lazaroff for the Environment News Service, "This research could force climate experts to recalculate the benefits of growing trees as a way to offset human caused emissions of carbon dioxide" (Lazaroff [August 8] 2002). Writing in *Nature*, Duke University ecologist Robert B. Jackson and colleagues concluded that in many locations, trees might be absorbing less carbon than soil emits once it is covered with grasses. "It had been proposed that the woody species might even increase soil carbon compared to the grasslands," explained Jackson, lead author of the *Nature* study. "People really didn't think that grasslands would store more carbon in the soil than woodlands" (Lazaroff [August 8] 2002). The rich black soils beneath many grasslands provide a long-term carbon repository, Jackson said.

> We found a clear negative relationship between precipitation and changes in soil organic carbon and nitrogen content when

grasslands were invaded by woody vegetation, with drier sites gaining, and wetter sites losing, soil organic carbon. Losses of soil organic carbon at the wetter sites were substantial enough to offset increases in plant biomass carbon, suggesting that current land-based assessments may overestimate carbon sinks. Assessments relying on carbon stored from woody plant invasions to balance emissions may therefore be incorrect. (Jackson et al. 2002, 623)

In general, the wetter a grassland, the greater its ability is to remove carbon vis-à-vis trees. This study indicates, according to Jackson, "As you move to increasingly wet environments, grasslands have a lot more soil carbon than shrublands and woodlands do. That was somewhat of a surprise. The analysis suggested that sites with the potential to store the most plant carbon also had the potential to lose the most soil organic carbon" (Lazaroff [August 8] 2002).

In a "News and Views" article accompanying the study published in the same issue of *Nature*, Christine L. Goodale and Eric A. Davidson of the Woods Hole Research Center in Massachusetts noted that the work of Jackson and colleagues will make measurement of carbon sinks' effects on climate change much more complicated. "Woodlands, savannas, shrublands and grasslands cover about 40 percent of the Earth's land surface, and so their potential role as carbon sinks . . . is a key factor in the global carbon budget," they wrote. "Measuring the effects of woody encroachment at particular sites is one challenge; extrapolating the results to regional or larger scales is quite another. Particular sites are certainly large sinks for carbon, but the global extent of grassland replacement by shrubland is highly uncertain" (Goodale and Davidson 2002, 594).

Jackson and his colleagues received support from another study indicating that agricultural lands may sequester more atmospheric carbon dioxide than forests and rivers. This study was conducted by researchers at the Yale School of Forestry and Environmental Studies and the Institute for Ecosystem Studies in Millbrook, New York. Carbon dioxide dissolved in rain and water in the soil acts as an acid, reacting with subterranean rocks to form dissolved carbonate alkalinity, which is then transported to the ocean. Peter Raymond, assistant professor of ecosystem ecology at Yale, and Jonathan Cole, an aquatic biologist at the Institute of Ecosystem Science, asserted that dissolved carbonate alkalinity emanating from the Mississippi River has increased dramatically during the past half-century. Cole, Raymond, and colleagues stated that the increase in dissolved alkalinity export is related to increases in precipitation that they document in the Mississippi watershed.

This research tends to contradict assumptions that reconverting agricultural fields to forests increases removal of atmospheric carbon dioxide by locking it into trees and soils. "Chemical weathering and the subsequent export of carbonate alkalinity from soils to rivers account for significant amounts of terrestrially sequestered carbon dioxide," Raymond and Cole wrote (2003, 88). They found that increases in alkalinity export from the Mississippi River during the previous fifty years were related not only to higher rainfall but also to the amount and type of land cover. "These observations have important implications for the potential management of carbon sequestration in the United States," they stated (p. 88).

Other research also suggests that planting forests to curb global warming could backfire. Thus, planting trees might not do much to curb global warming, contrary to assumptions expressed in the Kyoto Protocol. This assumption is deeply flawed, according to Richard Betts of Britain's Meteorological Office. "Carbon accounting alone will overestimate the contribution of afforestation to reducing climate warming," he told *New Scientist* ("Planting Northern Forests" 2001). Betts presented detailed calculations showing that planting trees across snow-covered swathes of Siberia and North America would heat the planet rather than cool it. On locations other than tundra, the cooling potential of forests is much less than previously supposed, Betts believes ("Planting Northern Forests" 2001). Where forests might replace snowy tundra, which usually reflects large amounts of solar radiation, heating of the Earth could accelerate. Betts calculates that at northern latitudes, warming as a result of planting forests would overwhelm any cooling effect resulting from the trees soaking up carbon dioxide.

Canada and Russia have proposed to plant forests in their empty tundras to help meet their Kyoto commitments, because a hectare of immature forest can absorb more than 100 tons of carbon each year, despite growing slowly ("Planting Northern Forests" 2001). Betts has calculated that the net warming effect of heat-absorbent forests in both regions is equivalent to an annual emission of 75 (English) tons of carbon per hectare. "I am not suggesting that we deforest," says Betts. "But afforestation is not always an effective alternative to cutting fossil fuel emissions" ("Planting Northern Forests" 2001).

William H. Schlesinger and John Lichter confirmed a view that forests have been overrated as carbon sinks:

> Such findings call into question the role of soils as long-term carbon sinks, and show the need for a better understanding of carbon cycling in forest soils. . . . Fast turnover times of organic carbon in the litter layer (of about three years) appear to constrain the potential size of the carbon sink. Given the observation that carbon accumulation in the deeper mineral soil layers was absent, we suggest that significant, long-term net carbon sequestration in forest soils is unlikely. (Schlesinger and Lichter 2001, 466)

HUMAN-INDUCED GREENHOUSE GAS EMISSIONS MIGHT OVERWHELM NATURAL CARBON SINKS

The oceans, soil, and vegetation now absorb about half of the carbon dioxide emitted into the atmosphere by human activity. Several studies project that, on balance, soil and vegetation will stop absorbing the gas and start emitting it by about 2050. The main causes will be greater respiration by plants in warmer soils and damage to the Amazonian rainforest caused by drier conditions. R. A. Gill and colleagues wrote, "The passive sequestration of carbon in soils may have been important historically, but the ability of soils to continue as sinks is limited" (Gill et al. 2002, 279). Gill's study of the reaction of a Texas grassland to a range of CO_2 levels showed that the availability of nitrogen in soil might limit the capacity of ecosystems to absorb increases in atmospheric carbon dioxide. The researchers said their study emphasizes the urgency with which the United States and other nations should adopt strict limits on

Officials pose for the formal photo at the prime ministerial residence of Kirribilli House during the inaugural ministerial meeting of the Asia-Pacific Partnership on Clean Development and Climate in Sydney on Wednesday January 11, 2006. © AP/Wide World Photos.

carbon dioxide emissions through diplomatic vehicles such as the Kyoto Protocol. ("Carbon Sinks" 2002).

"Based on fossil-fuel emissions, the carbon-dioxide concentration in the atmosphere should be going up twice as fast as it currently is," said Duke University's Robert B. Jackson. "However, natural systems such as the regrowing Eastern forests are currently taking up that extra carbon dioxide, so we're really getting a free ride now. . . . Many of us, myself included, believe that this free ride won't continue to the same extent that it has, because the incremental benefits of the extra CO_2 get smaller and smaller relative to other nutrient constraints" ("Carbon Sinks" 2002).

A large section of north Texas prairie was enclosed under two plastic-covered chambers. Researchers then exposed the grassland to varying concentrations of carbon dioxide. Over several growing seasons, the scientists conducted detailed biochemical and biological analyses of the grass plants and the soil and measured how the species composition of the plant community changed ("Carbon Sinks" 2002). "We found that many of the plants' physiological processes responded fairly linearly to increases in carbon dioxide, and plant production went up," said Jackson. "However, production and soil carbon storage basically saturated above 400 parts per million, a carbon-dioxide concentration very close to the current one [at about 380 ppm]" ("Carbon Sinks" 2002). The team found that as carbon dioxide concentrations rose, the plants used up much of the soil nitrogen, which limited further growth and carbon dioxide absorption.

"Considering the expected population increase, greater resource use per capita and the inability of natural systems to take up CO_2, we may well be looking at increases per year that are double what they are now, with atmospheric CO_2 concentrations as high as 800 parts per million in this century," Jackson concluded. "This means that the current lack of interest by the U.S. in participating in the Kyoto accords is especially unfortunate" ("Carbon Sinks" 2002).

Writing in *Science*, P. Falkowski and colleagues addressed the issue of weakening carbon sinks:

If our current understanding of the ocean carbon cycle are borne out, the sink strength of the oceans will weaken, leaving a larger fraction of anthropogenically produced CO_2 in the atmosphere or to be absorbed by terrestrial ecosystems. . . . As atmospheric CO_2 increases, terrestrial plant are a potential sink for anthropogenic carbon. . . . Because the saturation fluctuation decreases as CO_2 increases, terrestrial plants will become less of a sink for CO_2 in coming decades. . . . As in the case of marine ecosystems, we can predict that the negative feedback afforded by terrestrial ecosystems in removing anthropogenic CO_2 from [the] atmosphere will continue; however, the sink strength will almost certainly weaken. The exact magnitude of the change in sink strength remains unclear. (Falkowski et al. 2000, 293)

GLOBAL WARMING INFERRED FROM A DECREASE IN EARTH'S EMISSION OF RADIATION

Scientists from the Imperial College of London reported in the March 15, 2001, issue of *Nature* that the amount of long-wave, or infrared, radiation escaping from Earth into space had decreased over the previous three decades because an imbalance of the radiation is being retained near the surface by greenhouse gases. "These unique satellite spectrometer data collected 27 years apart show for the first time that real spectral differences [in long-wave radiation] have been observed and that they can be attributed to changes in greenhouse gases over a long time period," a team led by John E. Harries wrote (Davidson 2001, A-2).

The study was based on a comparison of radiation measurements by space satellites, including NASA's Nimbus 4 and a Japanese satellite. These satellites measured long-wave radiation from the atmosphere twenty-seven years apart, first during the early 1970s, then during the late 1990s. The scientists measured the long-wave radiation spectra corresponding to greenhouse gases, including carbon dioxide, methane, and chlorofluorocarbons. "We're absolutely sure, there's no ambiguity; this shows the greenhouse effect is operating and what we are seeing [the declining emission of terrestrial long-wave radiation] can only be due to the increase in the gases," Harries said (Davidson 2001, A-2). "Our results provide direct experimental evidence for a significant increase in the Earth's greenhouse effect that is consistent with concerns over radiative forcing of climate" (Harries et al. 2001, 355).

Proof that less long-range radiation is escaping the Earth does not equate to a certain degree of future warmth, according to Harries and colleagues, in part because greenhouse warming probably generates more cloud cover, which might exert a cooling effect. According to Harries, "Our next step is to assess whether these data can provide information about changes in not only the greenhouse gas forcing (heating of the atmosphere), but the cloud feedback, which is a response to that forcing." On the other hand, Tom Wigley of the National Center for Atmospheric Research stated, "The paper is an excellent piece of science, but it does not demonstrate global warming" (Davidson 2001, A-2).

TEMPERATURE READINGS IN THE EARTH'S CRUST

During the last half of the twentieth century, rocks in the Earth's continental crust have warmed along with the atmosphere and oceans, according to a team of Michigan and Canadian researchers. "Our findings remove any last doubt that this is anything other than a global phenomenon," said Henry Pollack, University of Michigan professor of geological sciences in a report by the Environment News Service ("Rock Measurements" 2002).

"Until recently, the story of global warming has been built up primarily on the basis of temperature measurements at the surface of the land and oceans," said Pollack. "These measurements have been painstakingly acquired and put together, and there has been enough information to reconstruct a temperature history for the Earth's surface for the past 140 years. But it's all based on surface measurements. . . . The magnitude of the warming we estimate is very similar to that which has come from the studies of the ocean, atmosphere, and ice. We believe it makes a persuasive case that the warming has been truly global" ("Rock Measurements" 2002).

According to the Environment News Service report, "The scientists based their analysis on temperature readings taken by lowering sensitive thermometers into holes drilled from Earth's surface into rock formations on six continents, including Africa, Asia, Europe, North America, South America and Australia." These readings revealed pervasive warming. The researchers' calculations, which were based on data from 616 bore holes, found evidence of an increase in the heat retained by the continents over the past 500 years, with more than half of that heat gain occurring during the twentieth century, and almost one-third of it since 1950 ("Rock Measurements" 2002).

TEMPERATURE MODELS: HOW MUCH WARMER?

Forecasts of future temperature rises resulting from increasing levels of greenhouse gases in the atmosphere reveal a great amount of uncertainty. English scientists early in 2005 published a study in *Nature* that

Aerial view of New York City. Courtesy of Corbis.

combined the results of 50,000 climate simulations—one so large that 90,000 private citizens in 150 countries lent their computers' power via the Internet to help conduct it. The headline-grabbing news of this study was its upper limit—the prediction of a temperature increase by the end of the twenty-first century of as much as 11.5 degrees C (about 21 degrees F), much higher than upper-limit forecasts by the Intergovernmental Panel on Climate Change (IPCC). This study also contained a very large range of possible outcomes, with its lower limit at 1.9 degrees C (about 3.5 degrees F) and the vast number of outcomes clustered in the 3 to 4 degree C range (Stainforth et al. 2005, 403).

The study, based on the global climate model of the British Meteorological Office's Hadley Centre for Climate Prediction and Research in Bracknell, United Kingdom, simulated expected results of a carbon dioxide level double that of the eighteenth century, at the onset of the industrial revolution. That level might be reached by 2050, with its effects manifest about fifty years later. The project is a collaboration of experts at Oxford and Reading universities, the Open University,

London School of Economics, Rutherford Appleton Laboratory, and the Hadley Centre. "When we started out we didn't expect anything like this," said Oxford University's David Stainforth, lead author of the study. "If this is the case, it's very dramatic and very scary" (Dayton 2005, 3). Stainforth said processing the results showed the Earth's climate is far more sensitive to increases in greenhouse gases than previously realized. He added, "An 11 C–warmed world would be a dramatically different world. . . . There would be large areas at higher latitudes that could be up to 20 degrees C warmer than today. . . . It is possible that even present levels of greenhouse gases maintained for long periods may lead to dangerous climate change. . . . When you start to look at these temperatures, I get very worried indeed" (Connor 2005, 10).

The IPCC's 2000 assessment said that while uncertainties remain, studies during the previous five years, including more-sophisticated computer modeling, indicated, that "there is now stronger evidence for a human influence" on the climate and more certainty that man–made greenhouse gases "have contributed substantially to the observed warming over the last 50 years" ("Pollution Adds" 2000). The assessment forecast that, if greenhouse emissions are not curtailed, the Earth's average surface temperatures might increase substantially more than previous IPCC estimates. The panel concluded that average global temperature increases ranging from 2.7 to as much as 11 degrees F can be expected by the end of the twenty-first century if current trends in emissions of heat-trapping gases continue. In its 1995 assessment, the panel had estimated increases of between 1.8 and 6.3 degrees F. The new assessment said higher temperature forecasts result mainly from more-sophisticated computer modeling and an expected decline in sulfate releases into the atmosphere, especially from power plants, for other environmental reasons. These sulfates tend to act as a cooling agent by reflecting the sun's radiation. "An increasing body of observations gives a collective picture of a warming world," the IPCC's 2000 assessment said ("Pollution Adds" 2000).

Climate scientists are designing more-sophisticated tools to simulate the carbon cycle, but the models, even today, are far from a match for nature's complex system. For the first time, in 2003, scientists incorporated several human and natural factors into a climate-projection model. The model used in the 2000 assessment anticipated that increased carbon dioxide in the atmosphere combined with a decrease in human-produced sulfates might cause accelerated global warming during the twenty-first century ("New Climate Model" 2003). Results of the study, completed by Chris D. Jones and colleagues at the Hadley Centre for Climate Prediction and Research, appeared in the journal *Geophysical Research Letters,* published by the American Geophysical Union (Jones et al. 2003).

The climate-carbon cycle experiment completed by Jones and colleagues is the first to take a more comprehensive Earth-systems approach to climate modeling. This "all-forcings experiment" incorporates all known influences on global warming, including carbon dioxide emissions, non–carbon dioxide greenhouse gases, human-produced sulfate aerosol levels, the reflection of solar radiation associated with sulphate in the atmosphere (the "albedo effect"), atmospheric ozone levels, levels of solar radiation, the effects of volcanic eruptions, and climate-carbon cycle feedbacks (Jones et al. 2003).

The all-forcings model indicated that anticipated reductions in human sulfate emissions would contribute to a reduction in their atmospheric cooling effect, resulting in a net warming. The model also forecasts that warming itself will enhance soil respiration, releasing carbon dioxide stored in soil into the atmosphere, causing atmospheric carbon dioxide levels to rise at an accelerating rate as the twenty-first century progresses. By the end of the century, the all-forcings model anticipates that this increase in carbon dioxide and decrease of sulfates would cause an average global temperature rise of 5.5 degrees C (9.9 degrees F) compared with 4 degrees C (7 degrees F) when these interactions are removed (Jones et al. 2003).

Other studies have reached similar conclusions. Researchers from the Hadley Centre for Climate Prediction and Research, for example, have forecast that land temperatures might rise by 6 degrees C during the twenty-first century (Houlder 2000, 2). This projection resulted from a computer model forecasting that global warming will accelerate as warmer conditions reduce the amount of carbon dioxide absorbed by soil and vegetation. This study is one of several that indicates that human society within a century will experience a temperature rise unprecedented in its history; by the twenty-second century, if fossil fuel use is not sharply curtailed, the Earth might be returning to the climatic days of the dinosaurs. "[Earth has] had warmer periods in the past," said Geoff Jenkins of the Hadley Centre, "but the rate of warming is quite unprecedented" (Radford 2000, 10).

Coupling (that is, combining) climate models with models of the carbon cycle, Peter M. Cox and colleagues found that "climate-cycle feedbacks could significantly accelerate climate change over the twenty-first century" (Cox et al. 2000, 184). This study supports the expectation that by the year 2050 the biosphere will have been exhausted as a carbon sink, as the ecosystem itself begins to add net carbon to the atmosphere. By that time, global warming is expected to reinforce itself. By 2100, according to Cox and his coauthors, emissions of carbon dioxide from land will become large enough to balance the ability of the oceans to absorb some of it, accelerating the rising level of carbon dioxide in the atmosphere. Cox and colleagues' models portend an average rise in global temperatures of about 5.5 degrees C by the year 2100 (compared to averages in 1860), which is at the high end of the range estimated by the IPCC. This is a global average, with increases over the oceans expected to be lower than those over land masses. The increase over land is expected by this projection to be about 8 degrees C.

According to atmospheric researchers at NASA's Goddard Institute for Space Studies in New York City, if releases of greenhouse gases and soot increase as expected, the mean global temperature should rise by about 3 degrees F in the next fifty years (Cooke 2002, A-34). The NASA researchers also said that if strong efforts are made to slow or reduce emissions of greenhouse gases, the rise in temperature could be cut in half, perhaps to less than 1 degree F. Some rise in global temperature is inevitable, however, because heat is expressed in the atmosphere a few decades after fossil fuels that cause it are burned. "Some continued global warming will occur, probably about 0.9 degrees F. even if the greenhouse gases in the air do not increase further," said climatologist James E. Hansen, the study's lead author. He added that warming could be limited to "much less than the worst-case scenario" if strong action is taken (p. A-34). "The model yields good agreement with observed global temperature change and heat storage in the ocean," Hansen and twenty-seven coauthors wrote (Hansen et al. 2002). The research supervised by Hansen was a collaborative effort among nineteen institutions, including several universities, private industries, other NASA offices, and other federal agencies.

Hansen warned that reducing carbon dioxide emissions will be difficult "because fossil fuels are central to our energy systems, and hence the economy. However, we have slowed the growth rate from the 4 percent per year that existed to the mid-1970s to 1 per cent [now]. If we could slow that further, to zero per cent, keeping fossil fuels at today's rate, we could keep the climate change small" (Cooke 2002, A-34). The Hansen team's climate model was based on a computerized simulation of the entire atmosphere, taking into account factors such as continuing gas and soot emissions, the influence of the sun, cloudiness, and the ocean as a "heat sink." The conclusions would be firmer, Hansen said, if the model had "better representation of the upper atmosphere, and better representation of the ocean. We are making progress on both" (p. A-34).

GLOBAL WARMING AND DROUGHT CYCLES

An intense four-year drought across much of the United States, southern Europe, the Mediterranean, and central and southwest Asia between 1999 and 2003 might have been a harbinger of prolonged globe-spanning droughts in the future, scientists reported in the journal *Science* early in 2003. The researchers said that the global scope, persistence, and severity of this drought also might reflect thirty years of gradual global warming. The most recent episode of drought, which devastated crops and reduced food supplies from the southeastern United States to Afghanistan, partially ebbed by 2003, but "the researchers reported . . . that climate models predict an 'increased risk' of similar events in future years" (Toner 2003, 4-A). Notably, by 2004, the

A parched cornfield in Plant City, Florida, observed as the area was enduring its worst drought on record in more than 100 years. © Getty Images.

eastern United States had swung from drought to deluge conditions aided, during the late summer and early autumn, by the remains of four major hurricanes that pummeled Florida.

Shifting areas of relatively warm and cool water in the Pacific Ocean have set the stage for droughts and other climatic changes before, but Martin Hoerling and Arun Kumar of the National Oceanic and Atmospheric Administration wrote that the most recent episode was unlike anything in recorded history. "The state of the tropical ocean during 1998 to 2000 combined a naturally occurring cooling of the eastern Pacific with a less frequent, possibly inexorable warming of the Indian and western Pacific oceans," Hoerling explained (Toner 2003, 4-A). "It is considered likely," wrote Hoerling and Kumar, "that increased greenhouse gases intensified the global hydrological cycle in the later half of the twentieth century, a process that would render some regions wetter and others drier and would also increase the likelihood for extreme precipitation.... The warmth of the Indian and west Pacific oceans was unprecedented and consistent with greenhouse-gas forcing" (Hoerling and Kumar 2003, 691).

Warming tends to accelerate the usual atmospheric processes by which air rises in the tropics (increasing rainfall) and subsides in the subtropics (decreasing rainfall). If one looks at a map of the world, major deserts usually concentrate in the subtropics, with the wettest areas usually nearest the equator. Warmer temperatures accelerate both tendencies. These are general tendencies; they are not always consistent based on latitude. The western United States (except for the northwest coast), for example, is markedly drier than the East because the East has access to moisture from the Gulf of Mexico. The foothills of the Himalayas also are very wet, a result of orographic lift during the wettest months of the Indian monsoon.

Hoerling and Kumar asserted the combination of La Nina's cooler-than-average waters in the eastern tropical Pacific and persistently warm water in the western tropical Pacific created the "perfect ocean" for an atmospheric circulation pattern that produced a globe-girdling drought across a wide swath of the Northern Hemisphere (Hoerling and Kumar 2003, 691). "An almost unbroken zonal belt of high pressure wrapped the middle latitudes," Hoerling said (Toner 2003, 4-A). Some drought-stricken areas, which stretched from New England to Pakistan, received as little as half their average precipitation during the four-year period, according to the study.

Hoerling and colleagues studied actual climatic behavior and then tested it against three different models in three laboratories. Each time, they examined sea-surface temperatures between 1998 and 2002. The researchers were looking for conditions that would support intense, widespread drought (Kerr 2003, 636). A total of fifty-one simulations produced strikingly similar results, anticipating less rain in the United States, southern Europe, and southwestern Asia. "The modeling results offer compelling evidence that the widespread midlatitude drought was strongly determined by the tropical oceans," Hoerling and colleagues wrote. "It is thus more than figurative, although not definitive, to claim this ocean was 'perfect' for drought" (Radford 2003, 18).

The researchers expressed concern that a generally warmer world could cause such patterns to appear more often and persist longer. Warmth in the Indo-western Pacific could become a persistent pattern.

The area has warmed on an irregular basis since the 1970s, and might continue to do so. "There's a strong suspicion that the Indo-western Pacific warming trend is related to the global-warming trend," said meteorologist Mathew Barlow of Atmospheric and Environmental Research Inc. in Lexington, Massachusetts (Kerr [January 31] 2003, 636).

HEAT, GREEN PLANTS, AND RISING OZONE LEVELS

According to a team of scientists from Britain's York University, when temperatures top 95 degrees F (35 degrees C), some green plants release isoprene and turpene to cool themselves. These chemicals then react to produce ozone, which is particularly dangerous for children, the elderly, and asthmatics. Isoprene, released by deciduous trees, and turpene, manufactured by evergreen trees, helps the trees protect their leaves from heat as well as damage by the sun. Once released, the chemicals increase the rate at which sunlight breaks down nitrogen oxide into ozone (Alleyne and Fenton 2004, 3).

The team studied ozone levels in Chelmsford, Essex, during August 2003's record heat wave in Britain, during which temperatures reached 100 degrees F for the first time. "By chance, we picked the two weeks of the heat wave. What we discovered was startling," said Alastair Lewis, who led the research. "When the temperature reached the high 90s and topped 100, plants and trees, which normally give off relatively small amounts of isoprene, started to produce greatly increased amounts" (Alleyne and Fenton 2004, 3).

GLOBAL WARMING AND STRENGTH OF
THE ASIAN SOUTHWEST MONSOON

David M. Anderson and colleagues reconstructed wind speeds of the Asian southwest monsoon for the last 1,000 years, using fossil *Globigerina bulloides* (sediment), which exists in abundance in box cores from the Arabian Sea. They found that "monsoon wind speed increased during the past four centuries as the Northern hemisphere warmed" (Anderson, Overpeck, and Gupta 2002, 596). They inferred that "the observed link between Eurasian snow cover and the southwest monsoon persists on a centennial scale" (p. 596). Alternately, they wrote in *Science*, "The forcing implicated in the warming trend (volcanic aerosols, solar output, and greenhouse gases) might directly affect the monsoon. Either interpretation is consistent with the hypothesis that the southwest monsoon strength will increase during the coming century as greenhouse gases continue to rise and northern latitudes continue to warm" (p. 596).

The southwestern Asian monsoon affects nearly half the world's population; its effects range from the Sahara to Japan. In addition to greenhouse gases, the strength of the monsoon is governed by 10,000 to 100,000 year (or more) cycles in solar insolation related to the Earth's orbit around the Sun. The strength of the monsoon reached a 10,000-year low about CE 1600, coinciding with the Maunder Minimum, a period of reduced solar activity, as well as cold temperatures during the "Little Ice Age" (Black 2002, 528). Evaluating the work of Anderson, et al., David M. Anderson, writing in *Science*, said, "This argument has critical implications in the face of global warming" (Black 2002, 528). Monsoon intensity was most notable during the twentieth century, as the pace of temperature increases has accelerated.

CLOUDS, EVAPORATION, AND GLOBAL WARMING

Warmth causes evaporation, which might affect patterns of cloud formation. Would additional clouds trap more heat or reduce warming by deflecting sunlight? Would areas with fewer clouds heat further because they receive more sunlight? Climate-change skeptic Richard Lindzen, who advocates the "Iris effect," has argued that increasing evaporation would allow heat to escape into space, effectively nullifying warming caused by rising carbon dioxide levels in the lower atmosphere.

According to satellite studies and computer models, increasing evaporation associated with warming might cause the tropics' canopy of clouds to shrink. Scientists at NASA's Langley Research Center report that clearer tropic skies will allow more sunlight to reach the Earth's surface. "The result is that the 'Iris effect' slightly warms the Earth instead of strongly cooling it," said Bruce Wielicki of NASA ("Clouds, but No Silver Lining" 2002; Wielicki et al. 2002, 841).

Robert J. Charlson and colleagues wrote in *Science*:

> Man-made aerosols have a strong influence on cloud albedo, with a global mean forcing estimated to be of the same order (but opposite in sign) as that of greenhouse gases, but the uncertainties associated with the aerosol forcing are large. Recent studies indicate that both the forcing and its magnitude may be even larger than anticipated ... making the largest uncertainty in estimating climate forcing even larger. (Charlson et al. 2001, 2025–2026)

Human activity, while increasing emissions of greenhouse gases, also alters ways by which some clouds form, perhaps causing them to screen the sun and partially cool the Earth by changing the size and density of water droplets. Researchers at the California Institute of Technology and the University of Washington have described their work along this line. "Almost no work is actually being done to model in detail how clouds respond to the polluted climate," said O. Brian Toon, an atmospheric scientist at the University of Colorado who was not involved in this study (Revkin 2001, F-4). Clouds exert their strongest cooling influence when they are dense and composed of small droplets. A cloud with a large number of droplets blocks more light. Many scientists have assumed that particles in the air serve as "seeds" on which water vapor can condense. This view, taken to its logical conclusion, would mean that clouds would not form in an atmosphere bereft of "condensation nuclei" provided by sea salt, dust, and pollution.

An analysis led by John H. Seinfeld, professor of chemical engineering at Caltech, and Charlson, of the University of Washington, asserted that certain acids and organic compounds can limit the growth of individual cloud droplets. "If each droplet is constrained from growing ... the same amount of condensing water ends up spread out among a higher number of smaller droplets. The result is a more reflective cloud," the study said (Revkin 2001, F-4). Nitric acid, a common byproduct of emissions from vehicles, is one of the strongest shapers of this process; others include several organic molecules pro-duced when fuels and wood are burned. "The work has been restricted to computer models, so the team did not try to calculate just how large the cooling effect of these other kinds of pollution might be relative to the warming effect of heat-trapping gases," the study stated (p. F-4).

"The more scientists look at the indirect effects of human emissions on clouds, the more convinced they are that the effects are large," said James E. Hansen (Revkin 2001, F-4).

REDUCED TROPICAL CLOUD COVER AND WARMING

Global warming reduced cloud cover over the tropics in the 1990s, NASA researchers have found. More sunlight entered the tropics and more heat escaped to space in the 1990s than during the 1980s because less cloud cover blocked incoming radiation and trapped outgoing heat, the researchers said after examining twenty-two years of satellite measurements. "Since clouds were thought to be the weakest link in predicting future climate change from greenhouse gases, these new re-sults are unsettling," said Bruce Wielicki of NASA's Langley Research Center. "It suggests that current climate models may, in fact, be more uncertain than we had thought. Climate change might be either larger or smaller than the current range of predictions" ("Warming Tropics" 2002).

The observations capture changes in the radiation budget—the balance between Earth's incoming and outgoing energy—that controls the planet's temperature and climate. A research group at the Goddard In-stitute for Space Studies developed a new method of comparing the satellite-observed changes to other meteorological data. "What it shows is remarkable," said Wielicki. "The rising and descending motions of air that cover the entire tropics, known as the Hadley and Walker circula-tion cells, appear to increase in strength from the 1980s to the 1990s. This suggests that the tropical heat engine increased its speed." The faster circulation reduced the amount of water vapor that is required for cloud formation in the upper troposphere over the most-northern and most-southern tropical areas. Less cloudiness formed, allowing more sunlight to enter and more heat to leave the tropics ("Warming Tropics" 2002).

Interactions between aerosols and clouds are important drivers of climate change but have been poorly understood, provoking much scientific study. The effects of biomass burning on clouds in the Amazon valley have come under particular scrutiny. One study (Koren et al. 2004) examines the suppression of boundary-level clouds (popularly called fair-weather cumulous) in smoky areas during the dry season. Another study (Andreae et al. 2004; Graf 2004) examines clouds that seem to emit smoke as a byproduct of biomass burning during the transition between dry and wet seasons. The transport of such pollution upward in the atmosphere might make thunderstorms more violent.

AEROSOLS AND CLIMATE CHANGE

The complexity of aerosols' effects on the atmosphere is a major problem in climate modeling. Writing in *Nature*, Meinrat O. Andreae and colleges sketched the complexity of the problem:

> All aerosol types (sulphates, organics, mineral dust, sea salts, and so on) intercept incoming sunlight, and reduce the energy flux arriving at the Earth's surface, thus producing a cooling. Some aerosols (for example, soot) absorb light and thereby warm the atmosphere, but also cool the surface. This warming of atmospheric levels also may reduce cloudiness, yielding another warming effect. In addition to these "direct" radiative effects, there are several "indirect," cloud-mediated effects of aerosols, which all result in cooling: more aerosols produce more, but smaller, droplets in a given cloud, making it more reflective. Smaller droplets are less likely to coalesce into raindrops, and thus the lifetime of clouds is extended, again increasing the Earth's albedo. Finally, modifications in rainfall generation change the thermodynamic processes in clouds, and consequently the dynamics of the atmospheric "heat

engine" that drives all of weather and climate. (Andreae, Jones, and Cox 2005, 1187)

The upshot of this complex mixture of effects enhances warming as aerosols are removed from the atmosphere by pollution reduction. The degree of effect on consequent warming is open to intense debate, however, because of the complexity of the problem.

Nobel Laureate Paul Crutzen and Swedish meteorologist Brent Bolin, a former chairman of the IPCC, said during a workshop in Berlin that a diminishing aerosol "parasol effect" in the atmosphere during the twenty-first century could contribute to warming that might exceed the IPCC's estimates. "It looks like the warming today may have been only about a quarter of what we would have ... without aerosols," Crutzen said ("Global Warming's Sooty Smokescreen" 2003). Scientists at the Berlin workshop speculated that a growing load of aerosols in the atmosphere had reduced warming by about 1.8 degrees C during the twentieth century, two to three times as much as previously believed, indicating that the lower atmosphere is more sensitive to greenhouse gases than most models indicate.

Human activities now pump nearly as many aerosols into the atmosphere as natural processes; these aerosols may be observed as a brownish haze that collects on the southern shores of Mt. Everest as well as dust that travels from heavily populated areas of Asia across the Pacific Ocean to North America. Urban haze also has been observed flowing over the Indian Ocean from cities in Asia. This blanket of pollution has been given a name: "the Asian brown cloud." According to V. Ramanathan and colleagues, "Anthropogenic sources contribute almost as much as natural sources to the global AOD [aerosol optical depth]" (Ramanathan et al. 2001, 2119).

On August 10, 2002, a team of international climatologists led by Professor Paul Crutzen, whose work on stratospheric ozone depletion won the 1995 Nobel Prize, said that the brown cloud comprises a 10-million-square-mile, three-kilometer-deep haze of man-made pollutants (mainly from burning wood and dung as well as fuel for vehicles and power generation) spreading across the most thickly populated parts of the Asian continent, blocking as much as 15 percent of the sunlight. This haze, when it reaches the stratosphere, can spread around the world in a matter of days.

V. Ramanathan of the Scripps Institution of Oceanography, who has researched the phenomenon for several years, said that it was not just an Asian problem. "We used to think that the human impact on climate was just global warming. Now we know it is more complex. The brown cloud shows that man's activities are making climate more unpredictable everywhere. Greenhouse gases like carbon dioxide are distributed uniformly, but the particulates in the brown cloud add to unpredictability worldwide" (Vidal 2002, 3). More than 200 scientists have taken part in the Indian Ocean experiment to study the haze.

The brown cloud has been detected obscuring the sky around the highest peaks in the Himalayas and as far as 1,000 kilometers downwind from major Indian urban areas. The particles absorb heat, which tends to intensify global warming. At the same time, however, warming is mitigated because the haze partially obscures the sun, by about 10 percent, causing some loss in agricultural productivity, notably of Asian rice crops. The scientists are concerned that pollutants in the brown cloud might disrupt India's life-giving annual monsoon.

The brown cloud is described as a "dynamic soup" of vehicle and industrial pollutants, carbon monoxide, and minute soot particles or fly ash from the regular burning of forests and wood used for cooking in millions of rural homes (Vidal 2002, 3). John Vidal described the cloud in the *London Guardian*:

> At its seasonal peak, usually in January, the soot in the cloud bounces back sunlight into the upper atmosphere, and prevents evaporation from the sea, leading to less rainfall. This, in turn, is thought to be affecting the monsoon rains which determine agriculture, and adversely affecting the health and livelihoods of up to three billion people throughout Asia. (Vidal 2002, 3)

"Some places will see more drying, others more rainfall. Greenhouse gases and aerosols may be acting in the same direction or may be opposing each other," said Ramanathan (Vidal 2002, 3). "It is now undisputed that air pollutants and their chemical products can be transported over many thousands of kilometers. We urgently need data on the sources of the pollution, especially for China and India since they are contributing the bulk of the emissions," said a United Nations report (p. 3).

These aerosols, according to Ramanathan and colleagues, "produce brighter clouds which are less efficient at releasing precipitation. These in turn lead to large reductions in the amount of solar irradiance reaching Earth's surface, a corresponding increase in solar heating of the atmosphere, changes in the atmospheric temperature structure, suppression of rainfall, and less efficient removal of pollutants" (Ramanathan et al. 2001, 2119). Increasing density of aerosols also might weaken the hydrological cycle, "which connects directly to availability and quality of fresh water, a major environmental issue of the 21st century" (p. 2119).

An experiment headquartered on the Maldives Islands in the Indian Ocean took six weeks, cost $25 million, and included scientists from fifteen countries, a research ship, and a C-130 military transport aircraft crammed with instruments (Fialka 2003, A-6). The Indian Ocean Experiment (INDOEX) used instruments on land and on aircraft together with measurements made by NASA's Clouds and Earth's Radiant Energy System (CERES) sensor as it flew overhead aboard the Tropical Rainfall Measuring Mission satellite. The experiment's objective was to help scientists understand to what extent human-produced aerosols might offset global warming.

The Indian subcontinent offered the architects of the INDOEX campaign an ideal setting for their field experiment. The region was chosen for its unique combination of meteorology, landscape (relatively flat plains framed by the towering Himalayan Mountains to the north and open ocean to the south), and the large southern Asian population (roughly 1.5 billion) with its growing economy. "Together, these features maximize the effects of aerosol pollution," Ramanathan explained ("New NASA Satellite" 2001). Because of human industry, including automobiles and factories, burning vegetation particles build up in the atmosphere, where they are blown southward over most of the tropical Indian Ocean. The Indo-Asian haze has covered an area larger than that of the United States. The INDOEX team found atmospheric particles of natural origin, such as trace amounts of sea salts and desert dust, but it also found that 75 percent of the aerosols over the region, including sulfates, nitrates, black carbon, and fly ash, resulted from human activities. Most natural aerosols scatter and reflect sunlight back to space, thereby making our planet brighter. However, human-produced black carbon aerosol absorbs more light than it reflects, thereby making our planet darker.

"Ultimately, we want to determine if our planet as a whole is getting brighter or darker," Ramanathan stated. "We could not answer that question until we could measure the sunlight reflected at the top of the atmosphere with an absolute accuracy of 1 percent. The CERES sensors provide that accuracy for the first time ever from a space-based sensor" ("New NASA Satellite" 2001).

"A large reduction of sunlight at the surface has implications for the hydrological cycle because of the close tie between heat and evaporation," Ramanathan said. "It could change the heating structure of the atmosphere and perturb the climate system in ways we don't understand now. We don't know, for example, how this might affect the monsoon season" ("New NASA Satellite" 2001).

SOOT: A "WILD CARD" IN GLOBAL WARMING

Atmospheric soot is more plentiful worldwide than most scientists had previously thought. It also is contributing to the rapid heating of the Earth's atmosphere, according to an increasing body of research. Microscopic carbon particles in air pollution have long been linked to respiratory ailments, but scientists are trying to improve their understanding of how smoke in the air interacts with sunlight and chemicals to influence global warming (Polakovic 2003, A-30).

Dirty snow containing even small amounts (measured in parts per billion) of soot might be responsible for as much as a quarter of recent temperature rises in polar regions, according to NASA research. James

Hansen and Larissa Nazarenko, climate specialists at the Goddard Institute for Space Studies, said that even small amounts of soot contained in fossil fuel effluvia (most notably diesel exhausts) absorb more sunlight and inhibit the reflection of light and its attendant heat. Soot also causes snow to melt more quickly, contributing to rising sea levels, Hansen and Nazarenko said in an article in the *Proceedings of the National Academy of Sciences* (Hansen and Nazarenko 2004, 423–428). Before this study, soot usually had not been factored into climate models projecting global warming's speed and scope.

Hansen and colleagues asserted that soot from diesel engines and burned wood should be reduced before it reaches the Earth's large snow packs. The "dirty snow effect," as Hansen calls it, could add a new wrinkle to the debate over global warming and its causes. "I think that this is an important climate [force] that has been overlooked," Hansen said. "I searched through the several thousand pages of past IPCC documents. It is never mentioned" (Nesmith [December 23] 2003, 3-A). "In the developing world, it is more a matter of replacing biofuels with cleaner fuels or burning them more cleanly. One of the most effective ways for them to clean up the air is electrification, so that fuels are burned at a power plant where it is easier to eliminate emissions." Hansen added, "The bottom line is that the technologies exist for both developed and developing countries to greatly reduce emissions, but there needs to be a concerted effort to do that" (p. 3-A).

Hansen and Nazarenko commented, "For a given forcing it [soot] is twice as effective as CO_2 in altering global surface air temperature. This indirect soot forcing may have contributed to global warming of the past century, including the trends toward early springs in the Northern Hemisphere, thinning arctic sea ice, and melting land ice and permafrost.... Reducing soot emissions, thus restoring snow albedos to pristine high levels, would have the double benefit of reducing global warming and raising the global temperature at which dangerous anthropogenic interference occurs" (Hansen and Nazarenko 2004, 423).

According to an account by Gary Polakovic in the *Los Angeles Times*, "Understanding airborne soot is critical to determining how Earth's climate has been altered since the Industrial Revolution and how it might change in the future. About 1 million tons of so-called black carbon is floating above the Earth. About 40 percent comes from fossil-fuel power plants and factories and diesel-powered vehicles in industrialized countries" (Polakovic 2003, A-30). The rest of the atmosphere's load of soot comes from smoke from forest burning or stoves using wood or dung to cook and heat homes in developing nations.

Additionally, the increasing load of soot over southern Asia alters monsoon precipitation patterns and influences climate around the world, including "large warming over northern Africa and cooling over the southern United States, all superimposed upon a more general global-scale warming" (Chameides and Bergin 2002, 2214). Most soot emissions result from incomplete combustion of biofuels in developing countries. China and the United States emit roughly equal volumes of soot.

Soot also has been identified as a major cause of global warming by a Stanford University study. Diesel-run vehicles emit soot through their exhaust pipes. Coal is another major source of soot pollution, along with kerosene, jet fuel, and wood. "Soot (black carbon) may be responsible for 15 to 30 percent of global warming, yet it's not even considered in any of the discussions about controlling climate change," said Stanford Professor Mark Z. Jacobson (Shwartz 2001).

Human beings produce most of the soot particles that pollute the atmosphere, observed Jacobson. "The exact forcing is affected by how black carbon is mixed with other aerosol constituents. . . . The warming effect from black carbon may nearly balance the net cooling effect of other anthropogenic aerosol constituents. The magnitude of the direct radiative forcing from black carbon itself exceeds that due to CH_4 [methane], suggesting that black carbon may be the second most important component of global warming after CO_2 in terms of direct forcing" (Jacobson 2001, 695). "Soot consists primarily of elemental carbon, and 90 per cent of it comes from the consumption of fossil fuels—particularly diesel fuel, coal, jet fuel, natural gas and kerosene—as well as the burning of wood and other biomass when land is cleared," said Jacobson (Shwartz 2001). A reduction in worldwide soot emissions, he maintains, could prove beneficial in slowing the pace of global warming.

Jacobson and Hansen's views are revisions of earlier, simpler assumptions that all sulfate aerosols exerted a negative (cooling) influence in the atmosphere. Jacobson found that "the darker the particles are, the more solar energy they can absorb, heating the atmosphere" (Andreae 2001, 671). Satellite photographs taken during 1999 showed massive plumes of black carbon obscuring the Indian Ocean west of the Indian subcontinent, which added "maybe as much as 25 percent to the previously known global sources of this pollutant" (pp. 671–672). Soot's lifetime in the atmosphere is very short, usually one to two weeks, however, so efforts to cut emissions could have a nearly immediate environmental payoff.

"Only a handful of studies have considered the impact of soot on global warming," Jacobson said. "Most of those were based on the premise that soot never mixes with other particles in the atmosphere" (Shwartz 2001). Scientists have known for many years, however, that floating soot particles combine with dust and chemicals in the air, noted Jacobson. This is a crucial point, he said, because mixtures containing black carbon absorb more sunlight than particles of pure black carbon. Therefore, soot in its mixed state has a potential to make a significant contribution to global warming (Shwartz 2001).

"In the past, researchers have felt that soot really didn't have a significant warming effect," said Michael Bergin, assistant professor of earth and atmospheric sciences at the Georgia Institute of Technology and coauthor of an article on soot in *Science*. "As we've learned more about the amount of black carbon emitted by countries like China and India, it appears that soot could have important climate effects, and that these effects may be almost as much as those of carbon dioxide" (Bowman 2002).

Simulations conducted by Jacobson and other researchers show that particles of pure soot are very likely to end up in mixtures containing dust, sea spray, sulfate, and other chemicals within five days after entering the atmosphere. These findings are consistent with several atmospheric field studies, including a 1999 survey that found that more than 93 percent of all soot above the North Atlantic Ocean contained particles of sulfate. Jacobson then programmed his computer to simulate how millions of tons of mixed soot would affect the Earth's climate.

Similar cutbacks in soot emissions could prove to be a very effective way to counter global warming, according to Jacobson. He pointed out that technologies exist or can be developed to remove excess soot produced in fireplaces, truck engines, and other sources of black carbon. "We can also make efforts to control biomass burning and reduce our reliance on soot-producing fuels, such as coal and diesel," he asserted (Shwartz 2001).

Another study of soot's effects drew data from AERONET, a network that provides data on aerosol optical depth from more than 250 sites worldwide. The authors described the prevalence of atmospheric soot in the *Proceedings of the National Academy of Sciences* (Sato et al. 2003, 6319). "The study indicates there is a lot more black carbon in the atmosphere than we thought," said Dorothy Koch, a coauthor of the paper and an atmospheric scientist at the Goddard Institute for Space Studies. "All black carbon does is absorb sunlight. If you put more into the atmosphere, you increase the warming" (Polakovic 2003, A-30). "Soot is the wild card," said Stanford University climatologist Stephen Schneider. "And this stuff can get pretty uncertain. It's very complex. It could have any number of effects." Among the outcomes, soot particles might help form clouds that reflect sunlight, or it might concentrate in clouds, heat them up, and evaporate them to contribute to global warming. Also, the effect depends on how high soot accumulates in the atmosphere and on the landscape below it, Schneider said (p. A-30).

Theodore Anderson and colleagues wrote in *Science* that while aerosols presently act as a damper on global warming, their effects could diminish in the future:

Although even the sign of the current total forcing is in question, the sign of the forcing by the middle of the 21st century will certainly be positive. The reason is that GHGs [greenhouse gases] accumulate in the atmosphere, whereas aerosols do not. Even if the most negative value of aerosol forcing shown in the figures turns out to be correct, the current range of plausible emissions scenarios indicates that GHG forcing will exceed aerosol forcing somewhere between 2030 and 2050. Thus, despite current uncertainties, forward calculations lead to the unambiguous conclusion that anthropogenic activity will inevitably result in a strong, positive forcing of Earth's climate system. (Anderson et al. 2003, 1104)

Surabi Menon of NASA and Columbia University joined with James Hansen and colleagues, writing in *Science* that their climate models associate soot from diesel engines and incomplete burning of fuels such as wood, crop residue, cow dung, and other sources, mainly in Asia, with several local climatic trends in China and India, including "increased summer floods in south China, increased drought in north China, and moderate cooling in China and India while the rest of the world has been warming [on average]. . . . We found precipitation and temperatures changes in the model that were comparable to those observed if the aerosols included a large proportion of absorbing black carbon (soot), similar to observed amounts. Absorbing aerosols heat the air, alter regional atmospheric stability and vertical motions, and affect the large-scale circulation and hydrological cycle with significant regional climate effects" (Menon et al. 2002, 2250). Unlike carbon dioxide emissions, which add to global warming by trapping heat in the atmosphere, soot emissions might contribute to global warming and climate change by absorbing sunlight, heating the air, and making the atmosphere more unstable, according to this study (Heilprin 2002).

The effects of soot in China also might be intensified by overfarming, overgrazing, and destruction of forests. In northern China, dust storms have become a regular climatic feature in Beijing and nearby areas, "with adhered toxic contaminants . . . [that are] cause for public-health concern" (Menon et al. 2002, 2250). Some residual dust from these storms has been detected as far away as North America. In addition, according to sources cited in this research, soot aerosols "are carcinogenic and . . . a major cause of deaths associated with particulate air pollution" (p. 2250).

"If our interpretation is correct, then reducing the amount of black carbon or soot may help diminish the intensity of floods in the south and droughts in the northern areas of China, in addition to having human health benefits," Hansen said. The research, based on data from Chinese ground stations provided by Yunfeng Luo of the Chinese Institute of Atmospheric Physics, might determine whether a similar pattern of disturbances exists in India (Heilprin 2002). Menon and Hansen used a climate computer model developed by NASA and pollution data from forty-six monitoring stations in China that was collected by Luo. They used computer models to run four simulations estimating the effects of soot on precipitation cycles over China. In addition to its effects in China, the computer simulation also showed "a pattern of larger warming over northern Africa and cooling over the southern United States from the soot, against a background of overall increases in global temperatures" (Bowman 2002).

THE ROLE OF JET CONTRAILS IN CLIMATE CHANGE

A study published in the *Journal of Climate* estimated that increasing coverage of cirrus clouds over the United States (to which air traffic is a major contribution) could increase tropospheric temperatures 0.2 to 0.3 degrees C per decade (Minnis et al. 2004, 1671). The study, by researchers at the NASA facility in Langley, Virginia, concluded that clouds from aircraft exhaust, or contrails, contributed to a warming trend of 0.27 degrees C per decade in the United States between 1975 and 1994. The NASA study is the first time weather observations were used to document temperature change relating to contrails, said Patrick Minnis, a senior research scientist at Langley. "Cirrus clouds can have a net warming or net cooling effect on the Earth, depending on how thick they are," Minnis said. Contrails can stretch to 1,600 kilometers and widen to 60 kilometers, depending on the weather. "Cirrus clouds from contrails tend to be thin, and the effect of thin clouds tends to be warming," he said (Schleck 2004, D-6).

Contrails from high-flying jets also might be helping to narrow the diurnal range of high and low temperatures, making days slightly cooler

Jet and contrails. Courtesy of Corbis.

and nights slightly warmer, according to measurements taken during the three days after the September 11, 2001, terrorist attacks on New York City and Washington, D.C., when nearly all air travel in the United States was grounded ("Contrails Linked" 2002). David Travis, a climatologist at the University of Wisconsin–Whitewater, led a study that offered some of the first evidence for the climate-changing effects of contrails. Travis and colleagues, including Pennsylvania State University geographer Andrew Carleton and University of Wisconsin–Whitewater undergraduate Ryan Lauritsen, used satellite images to compare cloud cover from those three days to thirty years of data for mid-September. Then they reviewed daytime and nighttime surface air temperatures across North America collected from 4,000 weather stations.

Results of this study indicated that jet contrails, like other forms of cloudiness, are compressing diurnal temperature ranges in certain regions of North America. "Scientists have been noticing unusual changes in diurnal temperatures for quite some time, but can't explain why," said Travis. "We're providing one possible explanation here. Maybe jet contrail coverage is one of the reasons for this shrinking temperature range" ("Contrails Linked" 2002; Travis, Carleton, and Lauritsen 2002, 593).

Travis said the findings of his study might complicate the global-warming debate because in some regions contrails offset some of the temperature increases anticipated by global-warming models. The study also underscores the point that not all influences on climate are global. Factors such as contrails can make a difference on a regional or local scale. The connection between contrails and temperatures was very difficult to measure before the terrorist attacks of September 11, 2001, and the resulting three-day shutdown of all commercial airline traffic. The satellite data from September 11–13, 2001, provided scientists with a view of nearly contrail-free skies for the first time in almost half a century.

THE JET STREAM AND MILDER WINTERS
IN THE NORTHERN HEMISPHERE

A team of atmospheric scientists has associated the general mildness of winters in the Northern Hemisphere during recent decades with variations in the Northern Hemisphere Annular Mode (NAMS), part of the high-level jet stream. When NAMS is in its "positive" phase, winds strengthen over the eastern Atlantic, bringing soggy gales to Britain while also keeping continental Arctic air locked to the north and east most of the time. The "negative" phase of NAMS brings fewer storms to the British Isles but allows more frequent intrusion of cold air.

A newspaper account (Henderson 2001) stated that NAMS, rather than global warming, is responsible for a string of recent mild winters in England, but the scientists quoted in the piece did not make this distinction. For example, David Thompson, an atmospheric scientist at Colorado State University (and one of the leaders of the cited research), was quoted as saying, "It is conceivable that the behavior of the Arctic Oscillation could be linked to the build-up of greenhouse gases in the atmosphere" (Henderson 2001). Thompson and Wallace's own words read: "If this trend proves to be anthropomorphic, as suggested by recent climate modeling experiments, statistics like those presented here may prove useful in making projections of what winters will be like later in this century" (2001, 88).

GREENHOUSE GASES AND WINTERTIME WARMING

Accumulation of greenhouse gases explains why the Northern Hemisphere is warming more quickly during winter months than the rest of the world during the last thirty years, according to a computer climate model developed by NASA scientists. They found that greenhouse gases, more than any of the other factor, increase the strength of polar winds that regulate Northern Hemisphere climate in winter. The polar winds that play a large role in the wintertime climate of the Northern Hemisphere blow in the stratosphere, eventually mixing with air and influencing weather close to the Earth's surface. The findings of Drew Shindell, Gavin Schmidt, and other atmospheric scientists from Columbia University and NASA's Goddard Institute for Space Studies appeared in the April 16, 2001, issue of the *Journal of Geophysical Research* (Shindell et al. 2001, 7193–7210).

Shindell and colleagues asserted that increases in greenhouse gases contribute to the persistence of stronger polar winds into the spring and contribute to a warmer early-spring climate in the Northern Hemisphere. A stronger wind circulation around the North Pole increases temperature difference between the pole and the midlatitudes. Shindell said that the Southern Hemisphere isn't affected by increasing greenhouse gases the same way, because it's colder and the polar wind circulation over the Antarctic is already very strong (O'Carroll 2001).

"Surface temperatures in the Northern Hemisphere have warmed during winter months as much as to 9 degrees F. during the last three decades, over 10 times more than the global annual average 0.7 degree Fahrenheit," said Shindell. "Warmer winters will also include more wet weather in Europe and western North America, with parts of western Europe the worst hit by storms coming off the Atlantic" (O'Carroll

2001). Year-to-year changes in the polar winds are quite large, according to Shindell. "But over the past 30 years, we have tended to see stronger winds and warming, indicative of continually increasing greenhouse gases" (O'Carroll 2001).

THE PROSPECTIVE ABRUPT NATURE OF CLIMATE CHANGE

Climatologists have been sharing the disquieting idea that small shifts in global conditions might lead to sudden and abrupt climate changes. The National Academy of Sciences has warned that global warming could trigger "large, abrupt and unwelcome" climatic changes that could severely affect ecosystems and human society (McFarling [December 12] 2001, A-30). "We need to deal with this because we are likely to be surprised," said Richard Alley, a Pennsylvania State University climate expert. "It's as if climate change were a light switch instead of a dimmer dial" (p. A-30). The report, which was commissioned by the U.S. Global Change Research Program, includes a plea for more research on the links between the land, oceans, and ice that might trigger abrupt change. Alley also suggests that many of today's models of climate change are too simple because they do not include such changes (p. A-30).

Alley and others have used gases within ice cores drilled from the Earth's ice sheets to produce a detailed record of the Earth's climate for the last 110,000 years. Graphed, this record often looks like a line of very sharp teeth. "We've gotten better and better records, and we've been able to say the changes were really big and really fast and affected a lot of the world at the same time" (McFarling [December 12] 2001, A-30).

"Although abrupt climate change can occur for many reasons, it is conceivable that human forcing of climate change is increasing the probability of large, abrupt events," wrote a team led by Alley (2003, 2005). At times, they wrote, regional temperature changes one-third to one-half as large as those associated with 100,000-year ice-age cycles have taken place within a decade (p. 2005). Increasing precipitation extremes—from deluge to drought and back again—also might be associated with abrupt temperature changes (see Chapter 7, "Drought and Deluge: Changes in the Hydrological Cycle"). An intense drought that played a major role in destroying the classic Mayan civilization may be an example of such a change (Alley et al. 2003, 2005–2006). Regional temperature changes of 8 degrees to 16 degrees C have been known, from the paleoclimatic record, to have occurred within a few years. Such changes have, in the past, been most likely during the beginning or end of ice ages (p. 2006).

Alley has described "threshold transitions" as being comparable to leaning over the side of a canoe: "Leaning slightly over the side of a canoe will cause only a small tilt, but leaning slightly more may roll you and the craft into the lake" (Alley et al. 2003, 2006). At just the right point, a small "forcing" might set into motion a very large climatic change. Thermohaline circulation and El Niño (ENSO) changes might be notable stress points in the global system, they assert.

Alley and colleagues concluded:

> If human activities are driving the climate system toward one of these thresholds, it will increase the likelihood of an abrupt climate change in the next hundred years or beyond. . . . We may see simultaneously both gradual and abrupt changes in floods and droughts. Abrupt changes are possible in ice sheets affecting sea level and ocean circulation, in permafrost affecting land-surface processes and greenhouse gas fluxes, and in sea ice and other parts of the climate system. (Alley et al. 2003, 2008)

The abrupt nature of climate change can be enhanced by several feedback mechanisms that compound the effects of any single forcing (see Chapter 3, "Feedback Loops: Global Warming's 'Compound Interest' "). One of the triggers of rapid climate change could be the conversion of solid methane hydrates (also called clathrates) below the oceans' surfaces to gaseous form, which has caused spikes of greenhouse warming several times in the Earth's climatic history (Kennett 2003). A given amount of methane is twenty-three times as powerful a greenhouse gas as carbon dioxide over 100 years, and sixty-three times as powerful over twenty years. James P. Kennett and colleagues argue that the methane "gun" also has triggered other feedback mechanisms that enhance atmospheric warming (Kennett et al. 2003, 147, 168).

Arctic snow and sea ice cover a large portion of the Earth with a white surface that reflects sunlight and heat back into space. If large parts of the Arctic ice cap melt, the darker liquid sea surface would accelerate warming. Oleg Anisimov, an expert on the cryosphere at the State Hydrological Institute in St. Petersburg, Russia, has asserted that such a change is already occurring. The snow and sea-ice cover in the Arctic has decreased 10 percent since the 1970s, and the ice has thinned markedly in that time. "Such changes are already enhancing the greenhouse effect," Anisimov said (McFarling [July 13] 2001, A-1).

The possibility of such abrupt changes complicates the task of policymakers in two ways. It could mean that the amount of time available to adjust to climate change will be much shorter than many government officials have believed. It also increases the uncertainty of predictions, indicating that future climate cannot simply be projected forward in a straight line from the present (McFarling [December 12] 2001, A-30). "We're a little spoiled by the last 30 years," said John M. Wallace, an atmospheric scientist at the University of Washington. "Many years, we're just barely breaking the previous record" (p. A-30).

THE SUN AS A MAJOR "DRIVER" OF CLIMATE CHANGE

Humankind's use of fossil fuels is only one influence on climate, although it becomes relatively more powerful as the atmosphere's overload of greenhouse gases grows. Another importance shaper of climate has been cycles initiated by the sun's generation of the energy that sustains all life on Earth. A team writing in *Nature* (Solanki et al. 2004, 1084–1086) found that the level of sunspot activity during the last two-thirds of the twentieth century was "exceptional," the highest in roughly 8,000 years. While the sun's cycles have had long-term effects on climate, these authors asserted that "solar variability is unlikely to have been the dominant cause of the strong warming during the past three decades."

In a paper published online by *Science* paleoceanographer Gerard Bond of the Lamont-Doherty Earth Observatory in Palisades, New York, and colleagues reported that the climate of the North Atlantic has warmed and cooled nine times in the past 12,000 years in step with the waxing and waning of the sun (Kerr 2001, 1431–1433). "It really looks like the sun has mattered to climate," said Richard Alley, cited in a report by Richard Kerr in *Science*. "The Bond, et al. data are sufficiently convincing that [solar variability] is now the leading hypothesis to explain the roughly 1,500-year oscillation of climate seen since the last ice age, including the Little Ice Age of the 17th century" (Kerr 2001, 1431). This cycle is now in a rising mode, and "could also add to the greenhouse warming of the next few centuries," according to the Kerr report (p. 1431). According to Alley, solar variations can "gain leverage on the atmosphere" by changing circulation patterns in the stratosphere, which then effects the lower atmosphere and, finally, ocean circulation, where they affect such climate drivers as the rate at which "deep water" forms in polar regions (p. 1432).

Bond and colleagues investigated the influence of the Sun on the climate of the North Atlantic from the Holocene, about 11,000 years ago, to the present, using data extracted from deep-sea sediment cores. They measured the residue of galactic cosmic rays, which are more intense at the sun's surface. This cycle (roughly 1,500 years) is measured against the coverage of ice in the subpolar North Atlantic. The authors, according to a critique of their work, "show that each of the expansions of cooler water that occur roughly every 1,500 years was associated with a strong minimum in solar activity. This remarkable result provides strong evidence that solar activity does indeed modulate climate on centennial to millennial time scales" (Haigh 2001, 2109). Bond and colleagues also suggested, according to Haigh, "That changes in ocean thermohaline circulation, associated with the supply of fresh, low-density water from the drift ice and coupled to atmospheric circulations, may amplify the direct effects of small variations in solar irradiance" (p. 2109).

Using lake sediment from southwestern Alaska, Feng Sheng Hu and colleagues have found indications that "small variations in solar irradiance induced cyclic changes in northern high-latitude environments" (Hu et al. 2003, 1890). These changes, on century-long time scales, were detected in Holocene climate, "similar between the subpolar regions of the North Atlantic and North Pacific, possibly because of sun-ocean-climate linkages" (p. 1890).

Solar activity measured by abundance of sunspots is slowing during the twenty-first century, after a hundred years of greater-than-usual intensity that had included a rising number of solar flares, sunspots and geomagnetic storms (Radowitz 2003). This change might partially counteract the effects of greenhouse forcing during the twenty-first century. During the twentieth century, the Sun is thought to have added between 4 percent and 20 percent of observed temperature increases (Radowitz 2003). British Atlantic Survey research suggests that in years to come the Sun will exert less of an impact on climate. Mark Clilverd, who led this study, said: "This work is speculative and relies on the idea that the Sun shows regular cycles of activity on time-scales of 10 to 10,000 years, and that its heat output and activity are related. . . . We believe the work is well-grounded and the effect of solar activity on Earth's environmental system will not increase in the way it has during the [twentieth] century" (Radowitz 2003). The reduction in solar activity might be temporary, however. Clilverd said he expected solar activity to return to twentieth-century levels by 2200.

A study by Swiss and German scientists also has suggested that increasing radiation from the sun might be responsible for part of recent warming. Sami Solanki, director of the Max Planck Institute for Solar System Research in Gottingen, Germany, who led the research, said: "The Sun has been at its strongest over the past 60 years and may now be affecting global temperatures. The Sun is in a changed state. It is brighter than it was a few hundred years ago and this brightening started relatively recently—in the last 100 to 150 years" (Leidig and Nikkhah 2004, 5). Solanki said that the brighter Sun and higher levels of greenhouse gases have contributed to recent sharp rises in the Earth's temperature. Determining the exact proportion of the warming caused by changes in solar activity is impossible with present knowledge, he said.

Solanki's research team measured records for several hundred years of sunspots, which are believed to have intensified the Sun's energy output. Sunspot activity has been increasing for the past 1,000 years, they found, which roughly conforms to changes in temperature records. Along with other research (cited above), these researchers found that the increase in sunspot activity was especially notable during the twentieth century. Gareth Jones, an English climate researcher, said that Solanki's findings were inconclusive because the study had not incorporated other potential climate-change factors. "The Sun's radiance may well have an impact on climate change but it needs to be looked at in conjunction with other factors such as greenhouse gases, sulphate aerosols and volcano activity," he said (Leidig and Nikkhah 2004, 5).

TEMPORAL ENDURANCE OF GLOBAL WARMING

Although uncertainty remains regarding the amount of warming, many scientists agree that the trend would continue for at least the next hundred years, even if fossil fuel consumption could be slashed sharply. Robert Dickinson of the Georgia Institute of Technology's School of Earth and Atmospheric Sciences presented evidence to support this assessment at the annual meeting of the American Association for the Advancement of Science in Boston on February 17, 2002 ("Global Warming Could Persist" 2002).

"Current climate models can indicate the general nature of climate change for the next 100 to 200 years," Dickinson said. "But the effects of carbon dioxide that have been released into the atmosphere from the burning of fossil fuels last for at least 100 years. That means that any reductions in carbon dioxide that are expected to be possible over this period will not result in . . . less global warming than we see today for at least a century" ("Global Warming Could Persist" 2002).

"Given enough time, there may be as many winners as losers. However, many of the losers will be very unhappy, such as people who live on islands that will be put under water," Dickinson said. "It will take a lot of time for humans to adjust their systems to these changes. The biggest problem is the speed at which global warming is occurring. The only way to stop the increase of carbon dioxide in the atmosphere is to reduce CO_2 emissions to 20 to 30 percent of today's levels," Dickinson concluded. "I believe we will eventually achieve that goal, but it will probably take 100 years" ("Global Warming Could Persist" 2002).

CHANGING LAND-USE PATTERNS MIGHT AGGRAVATE WARMING

The growth of cities and industrial-scale agriculture might be responsible for as much as half of recent rises in temperature across the United States, according to a study by scientists at the University of Maryland. Meteorologists Dr. Eugenia Kalnay and Dr. Ming Cai found evidence that a temperature increase of 0.13 degrees C (0.234 degrees F) over fifty years might be attributed to changes in land use ("Half U.S. Climate Warming" 2003).

Kalnay and Cai's research used records from 1,982 surface stations that were located below 500 meters elevation in the forty-eight contiguous U.S. states during fifty years (1950–1999), as well as trends based on data from satellite and weather balloons. They reported: "The most important anthropogenic influences on climate are the emission of greenhouse gases and changes in land use, such as urbanization and agriculture. But it has been difficult to separate these two influences because both tend to increase the daily mean surface temperature. . . . Our results suggest that half of the observed decrease in diurnal temperature range is due to urban and other land-use changes. Moreover, our estimate of 0.27 degrees C. mean surface warming per century due to land-use changes is at least twice as high as previous estimates based on urbanization alone" (Kalnay and Cai 2003, 528–529).

Comparison of urban and rural weather stations, without including agricultural effects, would underestimate the total impact of land use changes, Kalnay and Cai asserted. The urban heat-island effect actually takes place mostly at night, the two scientists wrote, "when buildings and streets release the solar heating absorbed during the day" ("Half

U.S. Climate Warming" 2003). During the day, at the time of maximum temperature, the urban effect is one of slight cooling from the effects of shading, aerosols, and thermal-inertia differences between city and country that presently are not well understood, they believe ("Half U.S. Climate Warming" 2003).

Urban sprawl, deforestation, and agricultural practices can alter temperatures and rainfall patterns in ways that sometimes augment the effects of increased greenhouse gases. "Our work suggests that the impacts of human-caused land cover changes on climate are at least as important, and quite possibly more important than those of carbon dioxide," said Roger Pielke, an atmospheric scientist at Colorado State University. "Through land-cover changes over the last 300 years, we may have already altered the climate more than would occur [from] the radiative effect of a doubling of carbon dioxide," added Pielke, lead author of a study published in the August 2002 issue of *Philosophical Transactions: Mathematical, Physical & Engineering Sciences*, a journal published by the Royal Society of London (Lazaroff [October 2] 2002; Pielke, 2002).

Pielke and colleagues asserted that if carbon dioxide emissions continue to rise at recent rates, the level of carbon dioxide in the atmosphere will double from preindustrial levels by 2050. At the same time, land-surface uses will continue to change. According to a report by the Environment News Service, "Different land surfaces influence how the sun's energy is distributed back to the atmosphere. For example, if a rainforest is removed and replaced with crops, there is less transpiration, or evaporation of water from leaves. Less transpiration leads to warmer temperatures in that area" (Lazaroff [October 2] 2002).

Land-use changes can produce temporary regional cooling as well as warming. If farmland is irrigated, for example, more water transpires from moist soils, which cools and moistens the atmosphere, changing local precipitation and cloudiness. Forests might influence the climate in more complicated ways than previously thought, the authors contend. According to the Environment News Service report, "In regions with heavy snowfall, reforestation or the growth of new forests would cause the land to reflect less sunlight, meaning that more heat would be absorbed. This could result in a net warming effect, even though the new trees would remove carbon dioxide from the atmosphere through photosynthesis during the growing season" (Lazaroff [October 2] 2002). Reforestation also could increase transpiration in an area, putting more water vapor into the air. Water vapor in the troposphere, the lowest and densest part of the Earth's atmosphere, is the biggest contributor to greenhouse gas warming, the researchers said.

Australian researchers believe they have found strong evidence that clearing of land might trigger major climatic changes. A Macquarie University team said that its findings indicate that the climate can respond suddenly and dramatically to centuries of environmental abuse. The researchers used one of Australia's most powerful supercomputers to model changing rainfall patterns after the mid-1970s in the southwestern corner of Australia. Some parts of the region have suffered declines of up to 20 percent in winter rainfall, threatening Perth's water supply. Research by Andy Pitman, a Macquarie University atmospheric scientist, Macquarie researchers Neil Holbrook and Gemma Narisma, and Pielke suggests that land clearing might be responsible for about half of the rain shortfall (Macey 2004, 2). Pitman said the results of the modeling closely reflected the rainfall changes observed since the mid-1970s, suggesting that land clearing might exert five or ten times the influence previously expected. "We didn't expect to find that," he said. "It scared the hell out of us" (Macey 2004, 2).

Forests, which are now cleared for farming, once slowed moist winds blowing in from the Indian Ocean, Pitman said. "This slowing of the atmosphere causes turbulence, which in turn generates rainfall. Without the tree cover, the water in the atmosphere flows across the landscape and is deposited elsewhere. Our results suggest, and observations indicate, that it's falling farther inland, outside of the catchments that provide the Perth water supply. It may be another argument why we shouldn't be logging old-growth forests" (Macey 2004, 2). These findings also showed that climatic changes could appear decades or centuries after humans began interfering with the environment. "It may be that when the effects of deforestation suddenly exceed a threshold the climate is likely to respond in a dramatic way," Pitman said (p. 2).

ONCE UPON A GREEN VENUS?

Planetary ecosystems are forever evolving, a fact that may bring cold comfort to students of the greenhouse effect who cast their eyes upon Venus, where catastrophic greenhouse effect warming has raised surface temperatures to a toasty 850 degrees F (464 degrees C). Some contemporary theories argue that Venus may once have experienced a climate much more like that of today's Earth, "complete with giant rivers, deep oceans and teeming with life" (Leake 2002, 11). Two British scientists believe that they have found some evidence that rivers the size of the Amazon once flowed for thousands of miles across the Venusian landscape, emptying into liquid-water seas. According to a report by Jonathan Leake

The surface of Venus near the Maat Mons, showing lava flows. Courtesy of NASA.

in the London *Sunday Times*, these scientists used radar images from a NASA probe to trace the river systems, deltas, and other features they say could have been created only by moving water (p. 11).

Because of its thick cloud cover and searing surface temperatures, no one on Earth knew much about Venusian topography until 1990, when NASA's *Magellan* probe used radar to penetrate the clouds and map the surface. *Magellan*'s images showed that the surface had been carved by large river-like channels that scientists at the time thought were caused by volcanic lava flows. The images were later reanalyzed by Adrian Jones, a planetary scientist at University College of London. He and colleagues used the latest computer technology and found that the channels were too long to have been created by lava (Leake 2002, 11).

Jones and colleagues said the findings suggested life on Venus could have evolved on a parallel with Earth's. "If the climate and temperature were right for water to flow, then they would have been right for life, too. It suggests life could once have existed there" (Leake 2002, 11). Studies compiled by David Grinspoon of the Southwest Research Institute at Boulder, Colorado, suggest that Venus might have been habitable for as long as 2 billion years before an accelerated greenhouse effect dried its oceans.

Roughly twenty space exploration missions to Venus have returned enough data to construct an image of Venus today as a hellishly hot place, "its skies dominated by clouds of sulfuric acid, poisoned further by hundreds of huge volcanoes that belch lava and gases into an atmosphere lashed by constant hurricane winds" (Leake 2002, 11). Venus differs from Earth in one important respect: it has no tectonic plates that permit stresses to express themselves piecemeal, via earthquakes and volcanic eruptions. The scientists' research suggests that as recently as 500,000 years ago something (perhaps a surge of volcanic eruptions, which had long been contained by the lack of tectonics) triggered runaway global warming that destroyed the Venusian climate and eventually boiled away the oceans (p. 11). According to this research, warming on Venus might have been accelerated by heat released into its atmosphere by billions of tons of carbon dioxide from rocks and, possibly, vegetation. Today, Venus' atmosphere is mainly carbon dioxide.

3 FEEDBACK LOOPS: GLOBAL WARMING'S "COMPOUND INTEREST"

INTRODUCTION

Many climate scientists believe that the middle of the twenty-first century will include dramatic acceleration in global warming. In part, this acceleration will result from exhaustion of various natural "sinks" for greenhouse gases, which are described in the previous chapter. At about the same time, various feedback loops also are expected to accelerate increases in atmospheric greenhouse gas levels and, consequently, worldwide temperatures. These include several natural processes that add greenhouse gases to the atmosphere, such as melting permafrost in the Arctic and gasification of solid methane deposits (clathrates) in the oceans. In both cases, human-provoked warming caused by an overload of greenhouse gases in the atmosphere is expected to enhance the transformation of methane clathrates from solid into gaseous form, which will compound existing problems like a bank account drawing an environmentally dangerous form of compound interest. Evidence is accumulating that these processes already have begun. The danger, according to many people who are familiar with the paleoclimatic record, is this: once this journey has begun in earnest, any return trip may become a matter of many centuries as well as copious human pain and suffering.

Sir John Houghton, one of the world's leading experts on global warming, told the London *Independent*: "We are getting almost to the point of irreversible meltdown, and will pass it soon if we are not careful" (Lean 2004, 8).

The amount of time that temperatures and sea levels may continue to rise after carbon dioxide levels stabilize is not known with any degree of certainty. It is widely believed, however, that "the rise in mean global SAT [surface average temperature] and the world ocean level [due to thermal expansion] may...continue for several centuries after the stabilization of the carbon-dioxide concentration [in the atmosphere], due to the gigantic thermal inertia of the oceans.... The response of ice sheets to earlier climate changes may continue for several centuries after the climate stabilizes" (Kondratyev, Krapivin, and Varotsos 2004, 45).

In addition to possible increases in greenhouse gases from gasifying permafrost and clathrates, other feedback mechanisms are expected to enhance the effects of greenhouse warming during the twenty-first century. Two of the most important are an increase in amount of water vapor (itself a greenhouse gas) and changes in the planet's albedo, or reflectivity, as formerly ice-covered land and sea surfaces are replaced by darker-colored bare land and open water. A report by the National Academy of Sciences prepared during 2001 at the request of the George W. Bush White House estimated that in coming years only 40 percent of potential warming would be caused directly by elevated levels of greenhouse gases. The other 60 percent might be caused by feedback mechanisms. "Together," according to this report, "these two feedbacks amplify the simulated climate response to the greenhouse gas forcing by a factor of 2.5" (Victor 2004, 146). The same summary pointed out that this factor is an estimate and that different climate models forecast varying ranges of feedback responses, producing differing estimates of future warming.

RELEASE OF CARBON FROM THE ARCTIC ECOSYSTEM

Given the fact that one-third on the Earth's carbon is stored in far-northern latitudes (mainly in tundra and boreal forests) (Mack et al.

2004, 440), the speed with which warming of the ecosystem may re-lease this carbon to the atmosphere is vitally important to those making forecasts of global warming's pace and effects. The amount of carbon stored in Arctic ecosystems also comprises two-thirds of the amount presently found in the atmosphere (Loya and Grogan 2004, 406). Its release into the atmosphere will depend upon the pace of temperature rise—and the Arctic, according to several sources (see Part III: "Icemelt around the World") has been the most rapidly warming region of the Earth.

During 2004, Michelle C. Mack and colleagues presented results in *Nature* of a twenty-year fertilization experiment in Alaskan tundra during which "increased nutrient availability caused a net ecosystem loss almost 2,000 grams of carbon per square meter" (Mack et al. 2004, 440). While aboveground plant production more than doubled under warmer conditions, "losses of carbon and nitrogen from deep soil levels . . . were substantial and more than offset the increased carbon and nitrogen storage in plant biomass and litter" (p. 440). According to this study, increased releases of carbon to the atmosphere that are "primed" by increasing decomposition of organic matter could accelerate the rise in atmospheric carbon dioxide—and, therefore, warming.

By 2004, injection of carbon dioxide into the atmosphere from Arctic peat lands ceased to be merely a matter of speculation about the future. Svein Tveitdal, director of the United Nations Environmental Program's center in Norway, warned on February 7, 2001, that rising temperatures already were melting permafrost in areas of Scandinavia that the center monitors, which is adding greenhouse gases to the atmosphere. "Per-mafrost has acted as a carbon sink, locking away carbon and other greenhouse gases like methane for thousands of years," said Tveitdal. "There is now evidence that . . . the permafrost in some areas is begin-ning to give back its carbon. This could accelerate the greenhouse effect" (Lazaroff 2001). Tveitdal warned that temperature increases during this century of up to 10 degrees C could accelerate melting of permafrost, adding to global warming. Tveitdal's center, GRID Arendal, has pro-duced interactive maps illustrating the current extent of permafrost in blue as a baseline from which scientists and policymakers can track the melting and shrinking of the Arctic's icebound soils. "I do not think it is radical to say that the map will become progressively less blue in the coming years," he said (Lazaroff 2001).

Scientists analyzing stores of carbon dioxide heretofore locked up in peat lands have found that the gases were being released into the environment at an accelerating rate. They estimated that the carbon dioxide released from peat lands could exceed the amount produced by the worldwide combustion of fossil fuel as early as 2060 (Connor 2004, 9). If no action is taken and emission of greenhouse gases continues from peat lands, these scientists have speculated that no peat lands may remain by the end of the present century. Atmospheric levels of carbon dioxide will double, by their estimation, as gas that was locked into peat is released to the atmosphere.

In a series of experiments described in *Nature*, researchers found that the increase in atmospheric carbon dioxide observed in recent decades has had a direct impact on kick-starting the release of carbon from peat (Freeman et al. 2004, 195). "Under elevated carbon-dioxide levels, the proportion of dissolved organic carbon derived from recently assimilated carbon dioxide was ten times higher than that of controlled cases," they wrote. "Concentrations of dissolved organic carbon appear far more sensitive to environmental drivers that affect net primary productivity than those affecting decomposition alone" (p. 195). This research is the first to measure and forecast the release of carbon dioxide from peat bogs.

This research also was among the first direct physical evidence of a "positive feedback" between carbon dioxide in the atmosphere and the huge stores of carbon locked up on land, with an increase in one causing a corresponding increase in the other. Chris Freeman of the University of Wales in Bangor, who led the research team, said, "We've got an enormous carbon store locked up in peat bogs which is equivalent to the entire store of carbon in the atmosphere and yet this store on land appears to have sprung a leak" (Connor 2004, 9). The amount of carbon dioxide being released from peat lands is accelerating roughly at a rate of 6 percent per year, according to the research of Freeman and colleagues. "By 2060 we could see more carbon dioxide being released into the atmosphere than is being released by burning fossil fuel," he said (p. 9).

Tests on peat samples taken from three different sites in Britain show that increasing the amount of carbon dioxide in the air around the samples causes the peat itself to emit up to ten times the amount of carbon dioxide that it would have under usual conditions. Freeman said that peat bogs release carbon in a dissolved organic form. Emissions from peat lands into surrounding rivers and streams have increased by between 65 and 90 percent in past six years (Connor 2004, 9). "The rate of acceleration suggests that we have disturbed something critical that controls the stability of the carbon cycle on our planet," he said. Dissolved organic carbon in rivers and watercourses also may react with the chlorine in water-treatment processes to produce potentially carcinogenic chemicals, Freeman said (p. 9). "We've known for some time that CO_2 levels have been rising and that these could cause global warming. But this new research has enormous implications because it shows that even without global warming, rising CO_2 can damage our environment," he added (p. 9).

Additional evidence that the Earth's colder regions already are releasing additional greenhouse gases into the air has been provided by M. L. Goulden and colleagues, who studied boreal forests and found "clear evidence that carbon dioxide locked into permafrost several hundred to 7,000 years ago is now being given off to the atmosphere as warming climate melts the permafrost" (Davis 2001, 270). These researchers investigated black spruce forest and found that carbon dioxide was being emitted into the atmosphere beneath a biologically active layer containing moss and tree roots (Goulden et al. 1998, 214). The same is true, in many cases, for methane. According to Neil Davis, author of *Permafrost: A Guide to Frozen Ground in Transition* (University of Alaska Press, 2001), this "relict" carbon dioxide:

> represents a massive source since it is estimated that the carbon dioxide contained in the seasonally and perennially frozen soils of boreal forests is 200 to 500 billion metric tons, enough if all released to increase the atmosphere's concentration of carbon

dioxide by 50 percent. Hence, it is possible that the release of carbon dioxide from melting permafrost during warming, or locking it into newly frozen soil during cooling, may accelerate climate change. (p. 270)

In the measured words of science: "We have observed a 65 per cent increase in the dissolved organic carbon (DOC) concentration in freshwater draining from upland catchments in the United Kingdom over the past 12 years. Here we show that rising temperatures may drive this process by stimulating the export of DOC from peat lands.... [This process] may increase substantially as a result of global warming" (Freeman, Evans, and Monteith 2001, 785). In simple English, a very dangerous carbonic feedback loop has been engaged.

Methane emissions from bogs in Sweden have been increasing, according to an international research team led by the GeoBiosphere Science Centre at Sweden's Lund University. The Abisko region in sub-Arctic Sweden, which they studied, has long-term records of climate, permafrost, and other environmental variables, which they say make comparisons possible (Christensen et al. 2004).

"In the present study, airborne infrared images were used to compare the distribution of vegetation in 1970 with that of 2000," states a report of the American Geophysical Union. "Dramatic changes were observed, and the scientists relate them to the climate warming and decreasing extent of permafrost that was observed over the same period" ("Swedish Bogs" 2004). At one site, in Stordalen, these researchers estimated an increase in methane emissions between 22 and 66 percent from 1970 to 2000, according to lead researcher Torben R. Christensen, who works at the GeoBiosphere Science Centre (Christensen et al. 2004; "Swedish Bogs" 2004).

THE METHANE "BURP" OR CLATHRATE GUN HYPOTHESIS: PAST AS PRECEDENT

Warming oceans could produce intense eruptions of methane from the sea floor, accelerating global warming caused by humankind's industries and transportation. About 10,000 billion tons of methane are stored beneath the ocean and on the continents. By comparison, the contribution of humans to the atmosphere's inventory of greenhouse gases via burning of fossil fuels has amounted to about 200 billion tons of carbon. If even a small portion of the oceans' stored methane were to escape into the atmosphere, greenhouse warming could greatly accelerate. Among scientists, this mechanism has come to be called the "methane burp," or the "clathrate gun hypothesis."

During past periods of rapid warming, methane in gaseous form has been released from the sea floor in intense eruptions. An explosive rise in temperatures on the order of about 8 degrees C during a few thousand years accompanied a methane release 55 million years ago, called the Palaeocene/Eocene Thermal Maximum. Evidence abounds that "just a small amount of warming could kick-start a positive feedback loop between hydrate release and further warming, sending global temperatures soaring" (Schiermeier 2003, 681). Following the temperature spike 55 million years ago, the Earth eventually recovered its temperature equilibrium—but it took roughly 100,000 to 200,000 years to do so.

This period has assumed a crucial position in climate studies because atmospheric concentrations of carbon dioxide are believed to have reached 800 to 1,000 parts per million (ppm) at its peak—the level that could become prevalent by the end of the twenty-first century at present rates of increase. The warming triggered 55 million years ago was quite spectacular. Scientists drilling in the Arctic have discovered that temperatures in the Arctic Ocean rose suddenly to as high as 68 degrees F (20 degrees C). "What no one expected was how much warmer it really was. That is a huge surprise," said Andy Kingdon of the British Geological Survey ("North Pole" 2004). Scientists discovered fossils of marine plants and animals that had died quickly because they could not cope with the surge in temperatures.

Scientists have known for some time about the massive release of methane about 55 million years ago because they were able to detect a distinct chemical signature in the geological record. Until recently, however, the cause of this spike in greenhouse gases had eluded them. A European team, led by Henrik Svensen at the University of Oslo, discovered ancient "escape conduits" on the floor of the Norwegian Sea. Svensen and several colleagues suggested that eruption of carbon-rich sedimentary strata from these vents under the present-day Norwegian Sea might have provided the trigger. The researchers suggested in *Nature* that such conduits were common throughout the northeastern Atlantic 55 million years ago, as the sea floor literally ripped apart between Greenland and northern Europe (Munro 2004, A-10). The European scientists suggested that upwellings of molten rock from deep in the Earth heated and cooked organic material in sea sediments, producing excessive amounts of methane gas that bubbled out and into the atmosphere. "Similar volcanic and metamorphic processes may explain climatic events associated with other large igneous provinces such as the Siberian Traps (about 250 million years ago)," they wrote in *Nature* (Svensen et al. 2004, 542). Believing that carbon dioxide release alone was insufficient to explain the magnitude of warming, Gabriel J. Bowen and colleagues have made a case that rapid increases in humidity (water vapor being a greenhouse gas) occurred at the same time (Bowen et al. 2004, 495).

Some scientists who have studied past releases of methane believe that they could provide a dramatic analogue for what might happen if human transport and industry continue to pump massive amounts of carbon dioxide into the air. The rise in temperature that has been associated with this event is, indeed, very close to what the Intergovernmental Panel on Climate Change projects (given "business as usual" conduct) during the lives of our great grandchildren. "The case is getting stronger that global warming is associated with large hydrocarbon releases, and here's an example where it seems to have happened in the past," said Roy Hyndman, a senior scientist at the Pacific Geoscience Centre of the Geological Survey of Canada (Munro 2004, A-10). "If you put a lot of methane or other hydrocarbons into the atmosphere you can get abrupt and very large global warming and that's what we're doing now," said Hyndman. "The amount we are putting in now is huge in the geological context, very large" (p. A-10).

In the past, these spikes in methane levels have been associated with increased emissions from terrestrial wetlands, but evidence is mounting that ocean waters' temperature rose enough to convert methane clathrates (or hydrates) to gas. Kennett and colleagues make a case that the conversion of clathrates was much more important in driving climate change than increases in wetland emissions. However, wrote Gerald R. Dickens, an earth scientist at Rice University in Houston, "There are ... two gaping problems: the authors provide no compelling evidence that methane released from the sea floor passed through the water column or that atmospheric methane initiated climate change. Until these major holes are plugged, their full hypothesis should rightfully be considered highly speculative" (Dickens 2003, 1017). Dickens added, "It's the first really nice evidence that hydrocarbons were coming out of the sea floor at this time" (Munro 2004, A-10). Dickens and Hyndman said that the warming spike 55 million years ago is an intriguing, if not yet completely solid, analogue for what humans are doing to the atmosphere today. "The question is, will it just warm steadily as we add more and more carbon, or is there some point where it will warm a lot? There is a danger of a sudden or catastrophic big change," said Hyndman (p. A-10).

Hot magma was delivered into organic-rich sediments throughout the North Atlantic at this time, initiating a massive hydrocarbon discharge from the sea floor. The composition of the expelled fluids and the timing of their release have not been explicitly defined, however, leading to questions about how many gigatons of carbon were released. Some calculations put that amount at about 3,000 gigatons, a "stupendous amount of hydrocarbons," considering that the entire world's conventional deposits of oil and gas today total about 5,000 gigatons (Dickens 2004, 514). While his explanation for the warming spike 55 million years ago needs refining, Dickens asserted that if cause and effect can be proved, these events might become "an intriguing but imperfect analogue of current fossil-fuel emissions" (p. 515).

"It is the first time geology has isolated an individual methane release event in the distant past," said Santo Bains of Oxford University's Department of Earth Sciences. "Now we can see just how it played its part in the global warming process of that era" (Keys 2001). Bains led geologists during three years of research through the badlands of

Wyoming, parts of Antarctica, and off Florida's east coast. "By studying that event, we may well be able to understand the effect of future global warming on the Arctic," said Euan Nisbet, a paleoclimatologist at Royal Holloway College, University of London.

Could the warming episode about 55 million years ago have been ignited by different methane releases in different parts of the world? T. L. Hudson and L. B. Magoon have advanced a theory that, 52 to 58 million years ago in the Gulf of Alaska, "large amounts of sediment, eroded from freshly uplifted mountains, were deposited in deep water off the 2,200-kilometer coast of Alaska" (Clift and Bice 2002, 130). The oceanic tectonic plate, subducting below the continental plate, was generating heat and producing hydrocarbons, including liquid methane, from organic matter in the sediment. According to Peter Clift and Karen Bice, writing in *Nature*, the methane may subsequently have turned to gas and bubbled into the atmosphere, "increasing greenhouse-gas concentrations [that] might [help to] explain the higher global temperatures 58 to 52 million years ago" (p. 130). The same two authors believe that while this process might have contributed significantly to a prolonged period of global warming at that time, it was probably too geographically isolated to have been the sole cause of worldwide warming (p. 130).

James P. Kennett led a team that examined periods of rapid warming that occurred together with rapid release of methane gas from hydrates in the oceans during the last few hundred-thousand years. They established the so-called "clathrate gun hypothesis" as valid science (Kennett et al. 2003). The late quaternary climate record contains several brief atmospheric warming events that can be associated with dramatic changes in ocean circulation and greenhouse gas levels in the atmosphere. These events also occurred in association with spikes in atmospheric methane levels.

Writing in *Science*, Kai-Uwe Hinrichs and colleagues Laura Hmelo and Sean Sylva of the Woods Hole Oceanographic Institution provided a direct link between methane reservoirs in coastal marine sediments and the global carbon cycle, which is an indicator of global warming and cooling (Hinrichs, Hmelo, and Sylva 2003, 1214–1217). Molecular fossils of methane-consuming bacteria found in the Santa Barbara Basin off California that were deposited during the last glacial period (70,000 to 12,000 years ago) indicate that large quantities of methane were emitted from the sea floor during warmer phases of the Earth's climate in the recent past. Preserved molecular remnants found by the Woods Hole team resulted from bacteria that fed exclusively on methane and indicate that large quantities of this powerful greenhouse gas were present in coastal waters off California. The team studied samples that were deposited 37,000 to 44,000 years ago.

"For the first time, we are able to clearly establish a connection between distinct isotopic depletions in forams and high concentrations of methane in the fossil record," said Hinrichs, an assistant scientist in the Woods Hole Institution's Geology and Geophysics Department. "The large amounts of methane presumably released during one event about 44,000 years ago suggest a mechanism different from those underlying the emissions at warmer periods, i.e. slow decomposition of methane hydrate triggered by warming of bottom waters. The sudden release of these enormous quantities of methane was probably caused by landslides and melting of the methane hydrate" (Hinrichs, Hmelo, and Sylva 2003, 1214–1217).

GREENHOUSE GASES AND FOREST FIRES

Forest fires have become a major example of a feedback loop that might accelerate global warming. According to some research, under some circumstances, wildfires might emit more carbon dioxide than humankind's contribution. What's more, many present-day computer simulations of climate change do not take fires' contributions into account.

Widespread wildfires during the summer of 2002 changed areas of the western United States from a carbon sink to a net carbon source as drought stunted tree growth, according to computer modeling studies of fires in Colorado conducted by a team of researchers from Colorado State University, the U.S. Geological Survey, and the National Center for Atmospheric Research. "We're using the western United States as a

An Indonesian resident attempted to extinguish a fire near his home in Tanjung Selatan near Samarinda in Indonesia's Kalimantan Province, March 23, 1998. © AFP/Getty Images.

case study area where climate and land use are interacting in several interesting ways," said National Center for Atmospheric Research senior scientist David Schimel. Western lands, particularly evergreen forests, represent about half of all U.S. carbon storage, he said ("Wildfires Add Carbon" 2002). "More carbon is freed from storage during droughts, not only because more dry vegetation burns, but also because plants deprived of water grow slower, absorbing and storing less carbon in their tissues ("Wildfires Add Carbon" 2002).

Another important case has been provided by Indonesian fires that polluted air over Southeast Asia during the El Niño years of 1997 and 1998. Roughly 60,000 kilometers of peat swamps, an area twice that of Belgium, dried up and burned in Indonesia during 1997 (Richardson 2002, 5). Susan Page of Britain's University of Leicester, together with colleagues in England, Germany, and Indonesia, analyzed satellite photos and data gathered on the ground to estimate how much of the fire area's living vegetation and peat deposits burned (Cowen 2002, 14; Page et al. 2002, 61–65). Robert Cowen of the *Christian Science Monitor* sketched the situation: "In Indonesia, nature and human activity had prepared a massive subsurface fuel reservoir. Tropical forests built up thick peat deposits as vegetation died and decayed over many centuries. Forest clearance and drainage for logging and farming have tended to dry the peat. Drought due to the 1997 El Niño was all that was needed to make the circumstances right for a sustained conflagration when forest-clearing fires were lit that year" (p. 14).

Page and colleagues explained the difficulty of calculating exactly how much carbon dioxide the fires emitted, but the totals were massive, especially when one adds to Indonesia's fires the many others that have burned around the globe, notably during North America's intense drought in 2002. The work of Page and colleagues has major implications for climate-change modeling because, as they wrote in *Nature*:

Tropical peat lands are one of the largest near-surface reserves of terrestrial organic carbon, and hence their stability has important implications for climate change. In their natural state, lowland tropical peat lands support a luxuriant growth of peat swamp forest overlying peat deposits up to 20 meters thick. Persistent environmental change—in particular, drainage and forest clearing— threatens their stability, and makes them susceptible to fire. This was demonstrated by the occurrence of widespread fires throughout the forested peat lands of Indonesia during the 1997 El Niño event. (Page et al. 2002, 61)

In Indonesia, layers of peat as thick as 20 meters (66 feet) cover an area of about 180,000 square kilometers (112,000 square miles) in Kalimantan (Borneo), Sumatra, and Papua New Guinea (Richardson 2002, 5). Page and colleagues used satellite images of a 2.5 million-hectare study area in central Kalimantan from before and after the 1997 fires. According to their estimates, about 32 percent of the area had burned, of which peat land accounted for 91.5 percent. An estimated

0.19 to 0.23 gigatons of carbon were released into the atmosphere through peat combustion, with a further 0.05 gigaton released from burning of the overlying vegetation. Extrapolating these estimates to Indonesia as a whole, the researchers estimated that between 0.81 and 2.57 gigatons of carbon were released to the atmosphere in 1997 as a result of burning peat and vegetation in Indonesia (Page et al. 2002, 61). According to the researchers, "This is equivalent to [between] 13 [and] 40 percent of the mean annual global carbon emissions from fossil fuels," which contributed measurably to the largest annual increase in atmospheric CO_2 concentration detected since records began in 1957 (p. 61).

Jack Rieley of the University of Nottingham in the United Kingdom also believes that burning peat in Borneo is a major factor in the rapid rise of atmospheric carbon dioxide levels. As farmers continue to clear the forests by burning, the bogs catch fire and release carbon for months afterwards. Suwido Limin, a biologist from the University of Palangka Raya in the Indonesian province of Central Kalimantan, told *New Scientist* late in 2004 that the fires have now returned after an earlier peak during an El Niño–provoked drought 1998. "During October [2004], the atmosphere around Palangka Raya has been covered in thick smoke, with visibility down to 100 meters. The schools have been shut and flights cancelled," said Limin (Pearce 2004).

The fires in Indonesia had other environmental effects as well. Iron fertilization of Indian Ocean waters resulting from the massive wildfires may have played a crucial role in producing a red tide of historic proportions that severely damaged coral reefs, according to Nerilie J. Abram and associates, writing in *Science*. Their findings "highlight tropical wildfires as an escalating threat to coastal marine ecosystems" (Abram et al. 2003, 952).

The spread of human populations is aggravating fire dangers around the world. The fires that ravaged much of Indonesia during 1997 and 1998 were caused, in part, by drought-provoked El Niño conditions. They were intensified, however, by other fires set by peasants hired by local capitalists desiring to open forest land for farming, grazing, and other forms of development. The fires were illegal under Indonesian law when they were set in protected areas—but not if they could be blamed on El Niño, a natural condition. At least twenty-nine companies later were indicted for setting illegal fires in Indonesia's rainforests (Glantz 2003, 196–197).

Page and colleagues pointed out that "in Indonesia, peat land fires are mostly anthropogenic, started by local (indigenous) and immigrant farmers as part of small-scale land clearance activities and, also, on a much larger scale, by private companies and government agencies as the principal tool for clearing forest before establishing crops" (Page et al. 2002, 61). During the unusually long El Niño dry season of 1997, many of these "managed" fires spread out of control, "consuming not only the surface vegetation but also the underlying peat and tree roots, contributing to the dense haze that blanketed a large part of Southeast Asia and causing both severe deterioration in air quality and health problems" (p. 61).

Commenting on Page's study in *Nature*, David Schimel and David Baker of the National Center for Atmospheric Research in Boulder, Colorado, noted that two other independent studies of atmospheric carbon dioxide concentrations during that time period supported the conclusion that the fires were a major contributor to atmospheric carbon dioxide levels. Schimel and Baker explained that computer climate simulations assume that processes that emit carbon dioxide and remove it from the atmosphere operate smoothly and continuously (Cowen 2002, 14; Schimel and Baker 2002, 29–30). Episodic events such as wildfires play havoc with such simulations.

At present, no climate modeler knows exactly how to take catastrophic events in small areas that release carbon dioxide that has been locked away in peat or other carbon and methane reservoirs and factor them into world-scale forecasts of greenhouse gas levels. Such events "can evidently have a huge impact on the global carbon balance," Schimel and Baker wrote (Cowen 2002, 14). During 1997, the growth rate of carbon dioxide in the atmosphere was double the usual rate, reaching its highest level on record to that time, in large part because of the Indonesian peat fires. Most of the carbon injected into the atmosphere during those fires resulted from burning of peat rather than combustion of trees (Schimel and Baker 2002, 29). Indonesia's peat is unusually dense and high in carbon content. Today's satellites can detect land-use patterns on a fifty-meter resolution; it is estimated that a thirty-meter resolution will be required to factor effects such as Indonesia's peat fires into global climate models. This is important because relatively small-scale events "can have an appreciable effect on the carbon cycle" (p. 29, 30).

WHEN WILL FEEDBACKS TAKE CONTROL?

How much time remains before critical feedbacks lock into place? In January 2005, a world task force of senior politicians, business leaders, and academics warned that the point of no return would be reached within a decade. The report, *Meeting the Climate Challenge: Recommendations of the International Climate Change Task Force*, was the first to place a figure on the length of time remaining before cascading feedbacks resulting from human-provoked climate change irretrievably commit the Earth to disastrous changes, including widespread agricultural failure, water shortages and major droughts, increased disease, sea-level rise, and the death of forests (Byers and Snowe 2005, 1).

The report asserted that the tipping point would occur when the average world temperature had increased 2 degrees C above the average prevailing in 1750, which was before the industrial revolution began. By 2005, temperatures already had risen an average of 0.8 degrees C. The report also asserted that the tipping point would occur when the atmospheric concentration of carbon dioxide passes 400 ppm. With the level at 379 ppm in 2004, rising at 2 ppm a year, that threshold was about a decade away in 2005. The report was assembled by the Institute for Public Policy Research in the United Kingdom, the Center for American Progress in the United States, and the Australia Institute. The group's chief scientific adviser was Rakendra Pachauri, chairman of the United Nations' Intergovernmental Panel on Climate Change (McCarthy 2005, 1).

The report concluded: "Above the 2-degree level, the risks of abrupt, accelerated, or runaway climate change also increase. The possibilities include reaching climatic tipping points leading, for example, to the loss of the West Antarctic and Greenland ice sheets (which, between them, could raise sea level more than 10 meters over the space of a few centuries), the shutdown of the thermohaline ocean circulation (and, with it, the Gulf Stream), and the transformation of the planet's forests and soils from a net sink of carbon to a net source of carbon" (McCarthy 2005, 1).

The science of warming feedbacks is not settled knowledge. In some areas, such as the role of organic decomposition in soils under warmer conditions, a robust debate continues. Knorr and colleagues have written in *Nature* that "the sensitivity of soil carbon to warming is a major uncertainty in projections of carbon-dioxide concentration and climate" (Knorr et al. 2005, 298), as their findings indicate that "the long-term positive feedback of soil decomposition in a warming world may be even stronger than predicted by global models" (p. 298). In the meantime, the very idea that organic soil decomposition is sensitive to temperature at all has been challenged by other scientists (Giardina and Ryan 2000, 858–861). For climate models, the solution of this debate is no small matter, because soils contain twice as much carbon as the atmosphere (Powlson 2005, 204), so the rate at which warming may accelerate exchange from one to the other is, and will continue to be, an important factor as other feedbacks add more greenhouse gases to the air that sustains us.

4 GLOBAL WARMING AND OZONE DEPLETION

SURFACE WARMING, STRATOSPHERIC COOLING, AND OZONE DEPLETION

Greenhouse gases warm the atmosphere and hold heat near the surface like a blanket. Deprived of emitted warmth, the stratosphere cools, aggravating depletion of ozone, which protects plants and animals from ultraviolet radiation. Chemical reactions that drive ozone depletion tend to accelerate as the stratosphere cools, retarding the restoration of ozone that was anticipated after the ban on chlorofluorocarbons (CFCs) under the Montreal Protocol, which was enacted during the late 1980s. As levels of greenhouse gases rise, the cooling of the middle and upper atmosphere is expected to continue, with attendant consequences for ozone depletion. Because of this relationship, problems with ozone depletion depend, in a fundamental way, on mitigation of greenhouse warming.

The energetic nature of ultraviolet-B (UV-B) radiation can break the bonds of DNA molecules. While plants and animals are generally able to repair damaged DNA, on occasion damaged DNA molecules can continue to replicate, leading to dangerous forms of skin cancer in humans. The probability that DNA can be damaged by ultraviolet radiation varies with wavelength, with rays of shorter wavelength being the most dangerous. "Fortunately," wrote Paul A. Newman, "at the wavelengths that easily damage DNA, ozone strongly absorbs UV and, at the longer wavelengths where ozone absorbs weakly, DNA damage is unlikely. But given a 10-percent decrease in ozone in the atmosphere, the amount of DNA-damaging UV would be expected to increase by about 22 percent" (Newman 1998).

The Antarctic ozone hole formed earlier and endured longer during September and October of 2000 than ever before—and by a significant amount. Figures from NASA satellite measurements showed that the hole covered an area of approximately 29 million square kilometers in early September, exceeding the previous record from 1998. These record sizes persisted for several days. Ozone levels, measured in Dobson units (DU), fell below 100 DU for the first time. The area cold enough to produce ozone depletion also grew by 10 to 20 percent more surface area than any other year. The ozone-depletion zone was coming closer to New Zealand, where usual springtime ozone levels average about 350 DU. During the New Zealand spring of 2000, ozone levels reached as low as 260 DU when atmospheric circulation patterns nudged the Antarctic zone northward. Scientists usually regard an area of the stratosphere as ozone-depleted when its DU level falls below 220.

During the beginning of September 2003, early indications were that the area of depleted ozone over Antarctica was approaching near-record size again. By the end of the month, the area of severely depleted ozone was the second largest on record, at about the size of North America.

Discovery of the "Ozone Hole"

During 1985, a team of scientists working with the British Antarctic Survey reported a startling decline in "column ozone values" above an observation station near Halley Bay (Farman, Gardiner, and Shanklin 1985, 207–210). Ozone depletion had been suspected, in theory, beginning in the early 1970s, and actual ozone densities had been declining over the Antarctic since 1977. The size of the decline in 1985 was a shocking surprise, however, because theorists had expected stratospheric ozone amounts to fall relatively evenly over the entire Earth.

Mario Molina and Sherwood Rowland, the first scientists to discover the ozone "hole," had expected a largely uniform decline of 1 to 5 percent (Rowland and Molina 1974, 810–812). Scientists at that time did not realize how ozone depletion related to temperature in the stratosphere. The seasonal variability of the decline was another surprise because existing theoretical models made no allowance for it. Ozone values over Antarctica tended to decline rapidly just as the sun was rising after winter. During the middle 1980s, the cause of dramatic falls in ozone density over the Antarctic was open to debate. Some scientists suspected variability in the Sun's radiational output, and others suspected changes in atmospheric circulation. A growing minority began to suspect CFCs. These chemicals were not yet proven as a cause when, in 1987, a majority of the world's national governments signed the Montreal Protocol to eliminate CFCs.

Definite proof of CFCs' role in ozone depletion developed shortly thereafter, as J. G. Anderson and colleagues implicated the chemistry of chlorine and explained a chain of chemical reactions (later broadened to include bromides as a bit player)—the "smoking gun" that explained why ozone depletion was so sharp and limited to specific geographic areas at a specific time of the year (Anderson, Brune, and Proffitt 1989, 11465). The temperature of the stratosphere became a key ingredient in the mix—the colder the stratosphere, the more active the chlorine chemistry that devoured ozone. By the year 2000, according to Maureen Christie, ozone depletion was "significantly affecting ozone levels throughout the Southern Hemisphere" (Christie 2001, 86).

Restoration of stratospheric ozone might even become more closely linked to greenhouse warming as temperatures continue to rise near the surface of the Earth. Guy P. Brasseur and colleagues modeled the response of the middle atmosphere to a doubling of carbon dioxide levels near the surface. Their models indicated, "A cooling of about 8 degrees Kelvin is predicted at 50 kilometers during summer. During winter, the temperature is reduced up to 14 degrees K at 60 kilometers in the polar region" (Brasseur et al. 2000, 16). Increasing levels of methane also add to this effect. In addition to its properties as a greenhouse gas, "methane oxidation leads to higher water and OH concentrations in the stratosphere and mesosphere, and hence to less ozone at these altitudes" (p. 16).

Global Warming and Ozone Depletion Coupled

The effect of global warming on ozone depletion is significant enough that the rate of depletion might not decrease even as levels of CFCs decline, according to Markus Rex of the Alfred Wegener Institute for Polar and Marine Research in Potsdam, Germany, and his colleagues. "I was surprised to see these results," said Drew Shindell, an atmospheric scientist at NASA's Goddard Institute for Space Studies. "We never suspected the [existing] models were this far out of whack" (Rex et al. 2004; Ball 2004).

Levels of atmospheric ozone have not recovered significantly following the banning of CFCs in the late 1980s, bearing out the relationship between surface greenhouse warming and stratospheric cooling. Measurements of stratospheric ozone above Arrival Heights, near Scott Base, Antarctica, reached 124 DU, the lowest level ever recorded, on September 30, 2000. For years after that, ozone depletion seems to have stabilized, improving in some years, deteriorating in others, depending on stratospheric weather conditions.

The decline in stratospheric ozone is striking when viewed on a graph with any sense of historical proportion. As little as a decade or two will do. Until the 1990s, in the Arctic, springtime ozone levels ranged around 500 DU. By the year 2001 they were averaging 200 to 300 DU; in the Antarctic, in the days before the "ozone hole" (about 1980), Dobson-unit values ranged from about 250 to 350; by the year 2000, they ranged from 100 to 200. During 2000, the ozone-depleted area over Antarctica grew, at its maximum extent, to a size equal to Africa. While the polar reaches of the Earth have been suffering the most dramatic declines in ozone density, ozone measurements over most of the planet also have declined roughly 15 percent since the middle 1980s.

A theoretical possibility that greenhouse forcing could hasten ozone depletion was suggested by numerical model results reported in the

November 19, 1992, issue of *Nature*. J. Austin, N. Butchart, and K. P. Shine, climate modelers from the British Meteorological Office and Reading University, ran a model simulating conditions in the stratosphere in a world where atmospheric carbon dioxide concentrations had been allowed to double, compared to preindustrial levels—a world in which we could be living within fifty years at present growth rates (Austin et al. 1992, 221–225).

Polar stratospheric clouds (PSCs) are not new. They have been described as "nacreous clouds resembling giant abalone shells floating in the sky" (Tolbert and Toon 2001, 61). These clouds form twenty kilometers above the ground during early spring and are sometimes called "mother-of-pearl clouds" because they shimmer. Some of these stratospheric clouds have been reported in Scandinavia for a century, and Edward Wilson noted them on Robert Falcon Scott's 1901 Antarctic expedition. Sometimes the clouds shine with green and orange shades at sunrise and sunset (p. 61). Polar stratospheric clouds remained largely an atmospheric curiosity until the discovery of widespread ozone depletion over the Antarctic during the middle 1980s. Scientists surmised that the ozone loss was occurring in the only place where the stratosphere was cold enough to produce these clouds. Rex and colleagues studied climatic conditions in the Arctic and found a surprisingly strong relationship between ozone loss and the number and density of PSCs that form in the stratosphere despite the fact that it is extremely dry. Results were reported in *Geophysical Research Letters* (Rex et al. 2004). By the end of 2001, Michael Proffitt, the World Meteorological Organization's senior scientific officer, said, "The area with temperatures low enough for polar stratospheric clouds that initiate rapid ozone destruction to form during October is double that found during any earlier five-year period" (Kirby 2000).

Daniel Kirk-Davidoff and colleagues wrote in *Geophysical Research Letters* that increasing coverage of polar stratospheric clouds "in a positive feedback loop" has been associated with dramatic polar warming at the surface during periods of high carbon dioxide levels in the Earth's history, notably the Eocene (38 to 55 million years ago) and the Cretaceous (65 to 135 million years ago). During these periods, land

and surface-ocean temperatures are believed to have been much higher in the polar regions than at present, while scientists' proxies indicate that tropical temperatures were similar to or only slightly higher than present times. Winters also were much warmer in continental interiors than at present. The growth of polar stratospheric clouds' coverage in recent decades may indicate that global warming will lead to a similar type of climate in the future, as tropical flora and fauna extended into much higher latitudes than they do now ("Editors' Choice" 2002, 401). Since most climate models do not account for the role of these clouds, Kirk-Davidoff and colleagues asserted that such models underestimate the potential of future greenhouse warming (Kirk-Davidoff et al. 2002, 14659).

Chemical reactions involving PSCs play an important role in depletion of ozone levels in the stratosphere over the Antarctic. Scientists are increasingly worried that cooler temperatures in the stratosphere over the Arctic might lead to increased formation of PSCs, hastening further ozone losses there. The clouds provide reaction surfaces for chemical reactions that convert CFCs and other chlorine-based compounds into a reactive form that destroys stratospheric ozone. The colder the stratosphere, the more readily PSCs form. The reactions in question also usually accelerate as temperatures fall. These reactions are very temperature-sensitive, with important consequences for Arctic ozone levels. A small cooling in the Arctic stratosphere might lead to large ozone losses. In some cases, a 5 degree C decline in temperatures can multiply ozone loss by ten times. Coverage of stratospheric clouds has been increasing steadily over the polar regions since at least the late 1960s. If Rex's findings and models prove correct, then all predictions about future ozone depletion may be underestimated, said Shindell (Ball 2004).

Coverage of PSCs seems related to rising levels of chlorinated molecules, including CFCs and other synthetic chemicals, notably "unprecedented concentrations of reactive chlorine in conjunction with severe ozone loss" (Tolbert and Toon 2001, 61). Ozone loss during the winter of 1999–2000 was exceptionally severe, with record cold stratospheric temperatures, abundant polar stratospheric cloudiness, and substantial

denitrification. A colder stratosphere is likely to provoke ozone depletion regardless of denitrification, although theoretical models suggest that
denitrification could increase expected ozone loses in a future, colder
Arctic by up to 30 percent (p. 62).

"ROCKS" IN THE STRATOSPHERE

As scientists probe the connections between surface warming and
atmospheric cooling, they find more potentially dangerous complica-
tions. For example, a team of atmospheric scientists has discovered large
particles inside stratospheric clouds over the Arctic that could further
delay the healing of the Earth's protective ozone layer. The team found
large, nitric acid—containing particles that could delay the recovery and
make the ozone layers over both poles more vulnerable to climate
change, said atmospheric chemist David Fahey of the National Oceanic
and Atmospheric Administration's office in Boulder, Colorado.

Each winter in the stratosphere over the poles, water and nitric acid
condense to form PSCs that unleash chlorine and bromine, which de-
grade ozone. Later in the winter, nitrogen compounds help shut down
the destruction. Fahey's team found previously unknown nitric acid particles that remove nitrogen, allowing the destruction to continue. They
nicknamed them "rocks" because they are hundreds of times bigger than
other particles in the clouds.

The rocks form during the polar winter, when temperatures in the
stratosphere decline to colder than −90 degrees C. If global-warming
forecasts become reality, the cooling of the stratosphere, compelled by
the retention of heat near the surface, might cause more rocks to form,
accelerating ozone depletion. "If it gets colder and you get more
'rocks,' the depletion period is going to last longer. The chlorine can
continue to eat ozone," said Paul Newman, an atmospheric physicist at
NASA's Goddard Space Flight Center in Maryland (Erickson 2001, 37-
A). "What he got is really outstanding," Newman said of the findings
by Fahey's team. "This mechanism that we now understand really will
help us be able to more precisely predict what's going to happen in the
future" (p. 37-A).

Fahey led a team of twenty-seven researchers that included scientists from the National Oceanic and Atmospheric Administration in Boulder, the University of Colorado, the National Center for Atmospheric Research in Boulder, and the University of Denver. The scientists described their findings in the February 9, 2001 edition of the journal *Science*. "It's a major puzzle piece in the process by which ozone comes to be destroyed," Fahey said of the discovery of the rocks, which was made during a January 2000 flight over the Arctic in an ER-2, NASA's version of the U-2 spy plane (Erickson 2001, 37-A). A machine on the aircraft that was measuring nitrogen-containing gases "coughed out what looked like disastrous noise" (Kerr [February 9] 2001, 962). The "noise" turned out to be very large particles (measured against other Arctic cloud mass) containing nitric acid (HNO_3), which had been previously unknown to science. The particles averaged 3,000 times the size of other atmospheric particles in the stratosphere.

These polar stratospheric cloud particles (known in shorthand form as PSC rocks) remove reactive nitrogen from the atmosphere that would otherwise "tie up chlorine and bromine in inactive, harmless forms" through denitrification (Kerr [February 9] 2001, 963). The rocks also "provide surfaces where chlorine and bromine can be liberated from their inactive forms to enter their ozone-destroying forms" (p. 963). Additionally, the PSC rocks' large size causes them to fall more quickly than other particles, removing even more nitrogen from the stratosphere. Given all these factors, the PSC rocks "have significant potential to denitrify the lower stratosphere" (Fahey et al. 2001, 1026).

Fahey and colleagues concluded:

Arctic ozone abundances will remain vulnerable to increased winter/spring loss in the coming decades as anthropogenic chlorine compounds are gradually removed from the atmosphere, particularly if rising concentrations of greenhouse gases induce cooling in the polar vortex and trends of increasing water vapor continue in the lower stratosphere. Both effects increase the extent of PSC formation and, thereby, denitrification and the lifetime of active chlorine. The role of denitrification in these future scenarios is likely quite important. (Fahey et al. 2001, 1030)

Fahey and his colleagues estimated that ozone depletion in the Arctic stratosphere might not reach its peak until the year 2070, even with a steady decline in chlorine levels.

Not everyone is as sanguine as Fahey about the future of stratospheric ozone. The subject is a matter of some rather intense debate. Sherwood Rowland of the University of California at Irvine, who shared the 1995 Nobel Prize for Chemistry for his part in the discovery that CFCs destroy stratospheric ozone, said that the effect of global warming on ozone depletion should be short-lived. "The [ozone depletion] story is approaching closure, and that's very satisfying," Rowland said (Schrope 2000, 627).

Alan O'Neill, a climate modeler at England's University of Reading said that record-breaking ozone holes in the year 2000 were not surprising and that the ozone holes should heal by the year 2050. He added, however, "Higher concentrations of greenhouse gases . . . could push that date back a few decades" (Schrope 2000, 627). O'Neill believes that ozone losses will peak about the year 2005 and then decline. Shindell said that even though scientists are beginning to understand how global warming could delay ozone-shield recovery, "The agreement to limit production [of CFCs] has been an unqualified success. The science was listened to, the policy-makers did something, and it actually worked" (p. 627).

Dangers of Arctic Ozone Depletion

Jonathan Shanklin of the British Antarctic Survey, one of the three scientists credited with discovering severe ozone depletion over Antarctica, has warned that global warming threatens to deplete stratospheric ozone over the Arctic in a manner similar to the ozone hole over the Antarctic. Shanklin told the British Broadcasting Corporation's Radio 4 program *Costing the Earth* that the hole could ultimately affect the United Kingdom, bathing it in higher levels of cancer-causing ultraviolet radiation. Shanklin also has been quoted as saying that the Arctic's ozone troubles are not over. He was quoted by Alex Kirby of Radio 4 as saying that the Earth's ozone layer is cooling, which makes its recovery more difficult:

> The atmosphere is changing, and one of the key changes is that the ozone layer [stratosphere] is getting colder. It's getting colder because of the greenhouse gases that are being liberated by all the emissions we have at the surface. And when it gets colder, particularly during the winter, we can get clouds actually forming in the ozone layer, and these clouds are the key factor. Chemistry can take place on them that activates the chlorine and makes it very much easier for it to destroy the ozone. We think that within the next 20 years we're likely to see an ozone hole perhaps as big as the present one over Antarctica, but over the North Pole. (Kirby [October 26] 2000)

Solar flares and frigid stratospheric temperatures during the winter of 2003–2004 provoked the worst depletion of ozone above the Arctic since records have been kept, according to a team of scientists reporting in the March 2, 2005, issue of *Geophysical Research Letters* (Randall et al. 2005). The team reported that levels of nitrous oxides up to four times any previously observed, agitated by solar activity, had combined with bitter cold (about −110 degrees F) to drive the depletion of ozone to levels 60 percent below anything observed (records reach to 1985). "I don't think we can be confident about whether or not we're seeing an ozone recovery or if we're attributing recovery to the correct causes," said Cora Randall, lead author of the report, who is an atmospheric scientist at the University of Colorado (Human and McGuire 2005, A-8).

While most of the area covered by the Antarctic ozone hole is uninhabited, a similar Arctic ozone-depletion zone could affect parts of Europe, Asia, and North America. In addition to severe ozone loses over Antarctica, stratospheric ozone levels also generally have been declining in the Arctic for several years.

In the last few decades, particularly in the 1990s, anthropogenic influences on the natural ozone layer have resulted in severe ozone depletion in polar regions. . . . During March 1997, stratospheric ozone in the Arctic reached all-time lows with losses due to halogen-related chemistry rivaling the Antarctic losses observed since the infamous 'ozone hole' was discovered in 1985. Three-dimensional, coupled physical-chemical models indicate that depletion of stratospheric ozone in the Arctic will increase in the future as climate change cools the stratosphere and stabilizes the polar vortex during the period when stratospheric chlorine and bromine from human-made halogen compounds will remain high (the next 20–30 years). (Bodhaine, Ellsworth, and Tatusko 2001)

The levels of protective ozone over most of Canada won't recover during the twenty-first century and probably will deteriorate during its second half, according to one scientific study. That study contradicted earlier, more optimistic forecasts that ozone levels around the world would begin to recover by midcentury, thanks to a ban on synthetic chlorine compounds that destroy ozone. "The more we know, the more we realize we don't know," said Jack McConnell, an atmospheric science professor at York University. "Lower ozone levels could mean jumps of up to 10 percent in some skin cancers, which now strike almost 60,000 Canadians a year. The research also found that ozone levels would be lowest in summer, the season of greatest danger. Ozone in the stratosphere—the layer 10 to 40 kilometers above Earth—screens out the ultraviolet rays linked to skin cancers (Calamai 2002, A-8).

Chemistry and Ignorance: A Bromide-Based Disaster?

Paul J. Crutzen has asserted that problems with the stratospheric ozone layer could have been much worse if chemists had developed substances based on bromine, which is 100 times as dangerous for ozone, atom to atom, compared with chlorine. "This brings up the nightmarish thought that if the chemical industry had developed organochlorine compounds instead of the CFCs—or, alternatively, if chlorine chemistry had behaved more like that of bromine—then without any preparedness, we would have faced a catastrophic ozone hole everywhere and in all seasons during the 1970s, probably before atmospheric chemists had developed the necessary knowledge to identify the problem" (Crutzen 2001, 10). Given the fact that no one seemed overly worried about this problem before 1974, wrote Crutzen, "We have been extremely lucky." This shows, he has written, "That we should always be on our guard for the potential consequences of the release of new products into the environment . . . for many years to come" (p. 10).

Crutzen emphasized the danger inherent in taking chances with the Earth's climate system without understanding its chemistry. The history of atmospheric chemistry during the last few decades, he said, has been one of surprises. "There may be more of these things around the corner," he said (McFarling 2001, A-1).

Multiple Links between Warming and Ozone Loss

Global warming and ozone loss are associated on several levels, illustrating the complexity of atmospheric chemistry. For example, increases in stratospheric UV-B radiation also tend to decrease productivity of phytoplankton, which absorb carbon dioxide. Thus, increasing UV-B radiation destroys a major sink for carbon dioxide. Increasing UV-B also increases the decomposition rate of nonliving organic matter, causing rising emissions of carbon dioxide via natural processes (Betsill 2003, 919–920).

Research by Australia's national scientific agency, the Commonwealth Scientific and Industrial Research Organization (CSIRO), examined relationships between stratospheric ozone levels and two key greenhouse gases, nitrous oxide and methane. Methane, molecule for molecule, is about twenty-three times more powerful than carbon dioxide in causing global warming but remains in the atmosphere only for roughly a dozen years. Nitrous oxide is 300 times more powerful than carbon dioxide in causing global warming, remaining active for 120 years. Nitrous oxide comes from sources that are difficult to control, such as agricultural soils and vehicle emissions, so a continued buildup in the atmosphere would also be difficult to control. As it slowly breaks down, nitrous oxide destroys ozone (Calamai 2002, A-8).

A CSIRO computer model anticipated that reductions in methane emissions might worsen conditions in the stratosphere and could drive the ozone levels in 2100 down to 9 percent below 1980 levels. Methane is a greenhouse gas, but it also provides protection against ultraviolet radiation because it produces ozone as it breaks down chemically. According to an account in the Toronto *Star*, "Current global climate change strategy focuses on pushing down methane levels while letting nitrous oxide levels soar. The result is a further drop in ozone levels" (Calamai 2002, A-8).

STRATOSPHERIC WATER VAPOR AND GLOBAL WARMING

Tropical wildfires and slash-and-burn agriculture have helped double the moisture content in the stratosphere over the last fifty years, concluded Steven Sherwood, assistant professor of geology and geophysics at Yale University, after examining satellite weather data. "In the stratosphere, there has been a cooling trend that is now believed to be contributing to milder winters in parts of the northern hemisphere," he said. "The cooling is caused as much by the increased humidity as by carbon dioxide" ("Biomass Burning" 2002).

Water vapor assessment by ground-based, balloon, aircraft, and satellite measurements shows a global stratospheric water vapor increase of as much as 2 ppm by volume during the last forty-five years, a 75 percent rise. Modeling studies by the University of Reading in England indicate that since 1980 the stratospheric water vapor increase has produced a surface temperature rise that is about half of that attributable to increased carbon dioxide alone.

"Higher humidity also helps catalyze the destruction of the ozone layer," said Sherwood ("Biomass Burning" 2002). Cooling in the stratosphere causes changes to the jet stream that produce milder winters in North America and Europe. By contrast, harsher winters might occur in the high latitudes of the Arctic. Sherwood said that about half of the increased humidity in the stratosphere has been attributed to methane oxidation. No one seems to know, however, what has caused the rest of the additional moisture. "More aerosols lead to smaller ice crystals and more water vapor entering the stratosphere," Sherwood explained. "Aerosols are smoke from burning. They fluctuate seasonally and geographically. Over decades there have been increases linked to population growth" ("Biomass Burning" 2002). Ozone experts in Canada and the United States said the Australian findings are a serious warning about the risks in trying to manipulate parts of the atmosphere in isolation. "You have to look at all these chemicals and see how they interact and evolve over time," said Tom McElroy, an ozone specialist with the Meteorological Service of Canada (Calamai 2002, A-8).

Other causes not directly related to human activity also might be increasing stratospheric moisture levels, according to Philip Mote, a University of Washington research scientist. "Half the increase [of water vapor] in the stratosphere can be traced to human-induced increases in methane, which turns into water vapor at high altitudes, but the other half is a mystery," said Mote. "Part of the increase must have occurred as a result of changes in the tropical tropopause, a region about 10 miles above the equator, that acts as a valve that allows air into the stratosphere" ("Most Serious" 2001). "A wetter and colder stratosphere means more polar stratospheric clouds, which contribute to the seasonal appearance of the ozone hole," said James Holton, University of Washington atmospheric sciences chairman and an expert on stratospheric water vapor. "These trends, if they continue, would extend the period when we have to be concerned about rapid ozone depletion" ("Most Serious" 2001).

HYDROFLOUROCARBONS (HFCS): SOLVING ONE PROBLEM, AGGRAVATING ANOTHER

Following the ban of CFC-laced refrigerant Freon, another chemical, hydroflourocarbon (HFC) was introduced as an environmentally friendly replacement that would not further imperil stratospheric ozone. Shortly thereafter, however, HFC proved to be a very potent greenhouse gas, with as much as 4,000 times the global-warming potential, molecule for molecule, of carbon dioxide. During the year 2000, Coca-Cola announced plans to phase out the use of HFCs in its cold-drink vending machines.

By the year 2000, HFCs were being widely used not only in air-conditioning and refrigeration systems but also in aerosol sprays that clean computers and cameras, among other devices. A sixteen-ounce spray can contains roughly the same amount of HFCs as an automobile's air-conditioning system. Additionally, while air-conditioning and refrigeration systems use the same amount of HFC repeatedly in an enclosed loop, a spray can shoots HFCs into the atmosphere with each use. According to the Aerosol Industry Association of Japan, 1,850 tons of HFCs were distributed in about 4.5 million cans in 2003, up considerably from 1,050 tons in 1995. An estimated 80 percent of these HFC sprays are used to blow away dust. The gas specifically used for this purpose—HFC134a—has a global warming potential 1,300 times that of carbon dioxide ("Spray Cans" 2004). The Kiko Network, a Japanese organization involved in mitigating global warming, estimated that one aerosol can is as damaging, in terms of carbon dioxide released, as leaving on a 21-inch TV set for four hours every day for twenty-two years ("Spray Cans" 2004).

5 WEATHER WARS: GLOBAL WARMING AND PUBLIC OPINION

INTRODUCTION: THE EVOLUTIONARY NATURE OF SCIENCE AND POLITICS

Global-warming science illustrates the dynamic nature of all knowledge. It is an evolving account of discovery, often beset by conflict and controversy. The relationship between warming of the atmosphere near the surface of the Earth and ozone depletion in the stratosphere was unknown a decade ago, for example. Any projections into the future are, by necessity, made on the current knowledge base, which is constantly changing. A study of the history of these ideas—of the evolving understanding of stratospheric ozone chemistry, for example, as well as global warming—will provide an idea of how risky projecting into the future from the current state of knowledge might be. What might we know in a hundred years?

The evolution of scientific knowledge is often beset by political controversy—and very few fields of science are more controversial than global warming, with its worldwide stakes and array of established interests formulating their own versions of "sound science." Scientists might propose, but politicians and the public will dispose, and only after long, heated debate. While the debaters react to their experience, greenhouse forcing provides the evidence for our eyes and ears several decades after fossil fuels have been burned.

Thus, the "weather wars" will continue.

BUSINESS AS USUAL IN THE UNITED STATES

Shortly after the dawn of the third millennium on the Christian calendar, George W. Bush's administration showed no sign of becoming terrorized by carbon dioxide.

"You're talking about a president who says that the jury is out on evolution, so what possible evidence would you need to muster to prove the existence of global warming [to him]?" said Robert F. Kennedy Jr., author of the new book *Crimes Against Nature* (2004). "We've got polar ice caps melting, glaciers disappearing all over the world, ocean levels rising, coral reefs dying. But these people are flat-Earthers" (Mieszkowski 2004). James E. Hansen, director of NASA's Goddard Institute for Space Studies, told the *New York Times* that, on the subject of climate change, the Bush administration was "picking and choosing information according to the answer they want to get" (Revkin 2004, F-1). "In my more than three decades in government, I have never seen anything approaching the degree to which information flow from scientists to the public has been screened and controlled as it is now," Hansen told a University of Iowa audience on October 26, 2004. "This process is in direct opposition to the most fundamental precepts of science," he said. "This, I believe, is a recipe for environmental disaster" (Schoffner 2004). Waiting another decade for a serious examination of climate change's effects, said Hansen, "is a colossal risk" (Hansen 2004). Hansen had traveled to Iowa at his own expense, as a government employee on leave—as a resident, he said, "of Kintnersville, Pennsylvania" (Hansen 2004).

On August 28, 2003, White House appointees in the U.S. Environmental Protection Agency ruled that carbon dioxide could not be regulated as a pollutant. The decision, reached after heat had killed several thousand people in Europe (see Part II: "The Weather Now—and in 2100"), reversed a 1998 Clinton administration position. It meant

Heavy freeway traffic. Courtesy of Corbis.

that the Bush administration would not use the Clean Air Act to reduce carbon dioxide emissions from power plants, cars, or other sources. In a report to the United Nations made public during 2002, the Environmental Protection Agency said, that total U.S. greenhouse-gas emissions were projected to increase by 43 percent between 2000 and 2020. By 2002, the 280 million citizens of the United States were emitting a greenhouse gas load equal to that of 2.6 billion people living in 151 poorer countries (Speth 2004, 61).

In the midst of the Bush administration's denial of global warming and the science supporting it, late in 2003, a Pentagon study on the effects of global warming was leaked to the press. The study, titled "An Abrupt Climate Change Scenario and Its Implications for United States National Security," attempted to forecast events within the next fifteen years. In the Pentagon study's hypothetical catastrophe, the United States, "turn[s] inward [and] effectively seeks to build a fortress around itself to preserve resources. Borders are strengthened to hold back starving immigrants." And later, "an ancient pattern reemerges: the eruption of desperate, all-out wars over food, water, and energy supplies" (Jurgensen 2004, D-1).

The report anticipated, among other events, that:

- By 2007, violent storms smash coastal barriers, rendering large parts of the Netherlands uninhabitable. Cities such as The Hague are abandoned.
- Between 2010 and 2020, Britain becomes colder and drier as its climate begins to resemble Siberia's.
- Access to water becomes a major battleground. The Nile, Danube, and Amazon were mentioned as being at high risk.
- Severe droughts affect the world's major breadbaskets, including the U.S. Midwest, where strong winds cause soil loss.
- A "significant drop" in the Earth's ability to sustain its present population becomes apparent within twenty years.
- Bangladesh becomes nearly uninhabitable because of rising sea levels that contaminate water supplies.
- Rich areas such as the United States and Europe become "virtual fortresses" to prevent millions of migrants from entering after they have been forced from land drowned by sea-level rise or rendered uninhabitable by rising temperatures and spreading deserts.
- Future wars are fought over the issues of basic survival rather than religion, ideology, or national honor.
- Deaths from war and famine run into the millions. (Buencamino 2004, 21)

The Pentagon report received copious publicity that pointed out the irony of its conclusions contrasted with the see-no-evil attitude toward climate change in the White House, which was being run, for the most part, by executives on loan from the fossil fuel industry, including Bush and his vice president, Dick Cheney.

As with the disaster film *The Day After Tomorrow* (see below), one leaked Pentagon study was worth more media mileage in the weather wars than a thousand carefully nuanced scientific studies. The Pentagon report cast climate change in terms of national security. Written by two members of a California think tank, the report was not meant as a forecast, but as a potential crisis-planning exercise (Jurgensen 2004, D-1).

Late in August 2004, the Bush White House seemed to have undertaken a dramatic reversal of its previous position on global warming, as a position paper conceded that emissions of carbon dioxide and other heat-trapping gases are the only likely explanation for global warming. Citing the "best possible scientific information," an administration official, James Mahoney, delivered a report to Congress that essentially reversed the previous White House position set out by President Bush, who earlier had refused to link carbon dioxide emissions to climate change (Younge 2004, 18). The report was endorsed by the secretaries of energy and commerce, as well as by Bush's science adviser. Bush himself seemed unaware of the change in position when questioned by the press, however. Asked why "the administration had changed its position on global warming's causes, Bush replied, 'Ah, we did? I don't think so' "("Bush Says" 2004).

The people of the United States of America have been the most profligate producers of greenhouse gases in the history of humankind. They have done this, in the main, without regard or regret for the future of the planet. Each U.S. citizen produces, per capita, five times the world average carbon dioxide load. One gallon of gasoline, for example, combines with oxygen in the atmosphere to produce propulsive energy and almost twenty pounds of waste carbon dioxide (Sorensen 2001, A-1). That gallon of gasoline might transport a garden-variety sport-utility vehicle ten to fifteen miles in an average U.S. city.

The fossil-fuel economy will not die for want of raw material any time soon. Fossil fuel resources on Earth in 2003 were estimated at 5,000 gigatons of carbon, vis-à-vis global consumption of six gigatons of carbon per year worldwide. Clearly, fossil fuels will not run out until long after their effluvia have raised world temperatures high enough to make life miserable for nearly everyone (Lackner 2003, 1677).

With the ascent of George W. Bush's administration in the year 2000, along with Republican majorities in the U.S. Congress, global warming came to be regarded in some powerful U.S. political circles as something of a bothersome joke. Witness, for example, Senator Jim Inhofe, Republican of Oklahoma and chair of the Senate Environment and Public Works Committee. From the Senate floor on July 28, 2003, Inhofe called global warming "a hoax" that is "predicated on fear rather than science," perpetuated by "environmental extremists." Inhofe reserved special spite for Hans Blix as a "ridiculous alarmist" ("Inhofe Calls" 2003). Blix, the United Nations' chief weapons inspector in Iraq, had said that global warming posed a greater threat to humankind than terrorism.

To Inhofe, by contrast, a warmer climate is a friend of humankind. "Numerous studies," he told the Senate, "have shown that global warming can actually be beneficial to mankind" ("Inhofe Calls" 2003). One might invite the senator to tell this to the polar bears, but polar bears don't vote in Oklahoma. Inhofe embraced the Republican climate change mantra du jour, "natural variability" (i.e., we can't do anything about warming, so gas up that SUV, grin, and bear it).

As Senator Inhofe spoke, the Bush administration announced a funding initiative of several million dollars to study "natural" causes of global warming. Inhofe and Bush called this "science," but many environmentalists called it fossil-fueled ideology. Inhofe was not alone. Asserting that "thirty years ago, other 'experts' theorized that continued combustion of fossil fuels would cause global cooling due to a buildup of greenhouse gases in the atmosphere," Chuck Hoffheiser, in a letter to the *Wall Street Journal*, concluded (naming no names) that "the 'global warming crowd' [is] promoting a treaty [the Kyoto Protocol] that is nothing more than a recipe for global socialism" (Hoffheiser 2001, A-23).

The skeptics often equate greenhouse gases with economic prosperity, assuming that any attempt to reduce their use will cost the United States economy dearly. This is a policy decision, however, not an ironclad law of economics. Worldwide emissions of carbon dioxide declined 0.5 percent in 1998, to 6.32 billion tons, for example, while the world economy grew by 2.5 percent, indicating that "economic expansion need not be a casualty of stricter environmental rules" (Yam 1999, 32). These figures were released by the World Wildlife Fund. Great Britain's greenhouse gas emissions fell 7 percent between 1990 and 2001, even as energy consumption increased 10.5 percent ("Acid Rain" 2003).

Inhofe's line of reasoning had been borrowed from the Bush White House. In June 2003, Bush took a detailed report on global warming from the U.S. Environmental Protection Agency and reduced it to one noncommittal paragraph. Thus, an administration larded with fossil fuel interests once more dismissed global warming as a nonissue. Satirist Molly Ivins wrote, at the time: "Think of the possibilities presented by this ingenious solution. Let's edit out AIDS and all problems with drugs, both legal and illegal. . . . We can do away with unemployment, the [medically] uninsured, heart disease, obesity, and the coming Social Security crunch. We could try editing out death and taxes." (Ivins 2003, 4-A). As the Bush administration was editing global warming off its cognitive map, Britain's Office of Science and Technology was warning that English coastal residents would face a thirtyfold increase in flood damage by the end of the century as a result of rising seas and increasingly violent winter storms aggravated, in part, by global warming (see Part IV: "Warming Seas" and Part II: "The Weather Now—and in 2100").

Inhofe's "sound science" was drawn principally from an article published in the June 2003 edition of a small scientific journal, *Climate Research*, and written by Willie Soon and Sallie Baliunas, two well-known climate-change skeptics. Their research, which was partially funded by the American Petroleum Institute, asserted that the twentieth century was not unusually warm compared with earlier centuries, notably the "medieval warm period" (see below). The paper's methodology drew "stinging rebukes" from many climate scientists even as it became an article of faith in the Bush White House (Regalado 2003, A-3). In late July, following copious attention, three editors of *Climate Research* (including its editor in chief, Hans von Storch) resigned in protest "over the journal's handling of the review process that had allowed the article into print. It was flawed and should not have been published" (p. A-3).

The point of view of Senator Inhofe and President Bush does not reflect that of the world consensus on global warming (see Chapter 2, "Scientific Research: The Issue's Complexity"). Ross Gelbspan, a former *Boston Globe* editor whose 1997 book, *The Heat is On*, detailed industry efforts to discredit climate-change science, said that the case for continuing (and accelerating) greenhouse forcing is the result of "the largest and most rigorously peer-reviewed scientific collaboration in history." Gelbspan added: "The contradictory statements of a tiny handful of discredited scientists, funded by big coal and big oil, represent a deliberate and extremely reckless campaign of deception and disinformation" (Nesmith [May 30] 2003).

GLOBAL WARMING AND PUBLIC OPINION

By the end of the twenty-first century, after cascading feedbacks have made global warming a no-doubts reality, sweltering future generations might look back with a sense of irritated amusement at the debates that filled the newspapers, airwaves, and Web pages at the turn of the millennium. In a century, a future generation might remark at our generation's collective lack of foresight and imagination. In our time, skeptics cast doubt on "belief" in global warming, as if it is a religious doctrine, while carbon dioxide and temperature curves escalate steadily.

In the United States to date, with a few notable exceptions, global warming has been back-pages fare. The exceptions have involved major newspapers with environmental and scientific correspondents—Mike Toner of the *Atlanta Journal-Constitution*, Usha Lee McFarling of the *Los Angeles Times*, Andrew Revkin (and others) at the *New York Times*. In Britain, however, global warming has been front-page news for years—even to the point of screaming headlines in the tabloids. In Britain, global warming receives some of the fear-factor attention reserved lately in the United States for international terrorists.

Reporters and editors at four major United States newspapers followed the journalistic custom of balance "at the expense of accurately reporting scientific understanding of the human contributions to global warming," according to an analysis that appeared in a report, *Global Environmental*

Change ("Top U.S. Newspapers" 2004). The study, "Balance as Bias: Global Warming and the U.S. Prestige Press," examined coverage of human contributions to global warming in the *New York Times*, the *Washington Post*, the *Los Angeles Times*, and the *Wall Street Journal* from 1988 to 2002.

"By giving equal time to opposing views, these newspapers significantly downplayed scientific understanding of the role humans play in global warming," said Maxwell T. Boykoff, a doctoral candidate in environmental studies at the University of California, Santa Cruz, who coauthored the paper with his brother, Jules M. Boykoff, a visiting assistant professor of politics at Whitman College. "We respect the need to represent multiple viewpoints, but when generally agreed-upon scientific findings are presented side-by-side with the viewpoints of a handful of skeptics, readers are poorly served," added Boykoff. "In this case, it contributed to public confusion and opened the door to political maneuvering" ("Top U.S. Newspapers" 2004).

THE DAY AFTER TOMORROW

While climate change has not been a big-ticket news subject in the United States, it has had its moments. Quite a debate, for example, erupted during mid-2004 over a disaster movie, *The Day After Tomorrow*, which dealt with climate change the way Hollywood deals with everything, by blowing it enormously and violently out of proportion. Hollywood does this with the wild West, space aliens, love affairs, and just about every other form of dramatic encounter. Why not climate change? Some wags pointed out that director Roland Emmerich already had staged destruction of some of the same New York City landmarks that he flooded and froze in *The Day After Tomorrow* in his 1996 blockbuster, *Independence Day*. He also ruined the same city's skyline (as well as the Brooklyn Bridge) in *Godzilla*.

In the movie, Dennis Quaid plays a scientist, Jack Hall, who is ignored by a complacent White House. "The real power in this White House," according to Robert B. Semple of the *New York Times*, "is not the well-meaning but vacant president, but a reactionary vice president who is meant, unmistakably, to be Dick Cheney and is played that way by Kenneth Welsh" (Semple 2004). As Hall warns of looming disaster, the White House complains that reductions of fossil fuel emissions will bankrupt the economy. Soon enough, verbal weather wars are replaced by climatic disaster as only Hollywood can cast it. Ocean temperatures in the North Atlantic drop 13 degrees C in a few minutes. Savage tornadoes rip Los Angeles. New York City is swept by a huge tidal wave, followed in a New York minute by a huge ice storm that flash-freezes nearly everything in sight, except some ill-tempered wolves that escape from the Bronx Zoo (Semple 2004).

The idea for the movie was sparked by a nonfiction book, *The Coming Global Superstorm* by Whitley Strieber and Art Bell, which Emmerich read while he was directing the Mel Gibson movie *The Patriot*. After reading the book, Emmerich began to research possible effects of global warming. Some people who wanted to see global warming debated tended to forgive the movie's scientific lapses and thanked its director for raising the salience of the issue at street level, something that had been sorely lacking since the attacks on the World Trade Center on September 11, 2001, had satiated the media's appetite for world-girdling bogeymen.

While *The Day After Tomorrow* did focus some attention on global warming (even though most of it depicted record cold), many of the events in the movie had a tenuous grounding in climate science. Hail larger than that depicted in the movie already had fallen in Nebraska, for example. Cold had interjected into worldwide warmth in the North Atlantic because of breakdown of the thermohaline circulation in the past, most recently about 19,000 years ago. However, while Richard B. Alley and other glaciologists have found that climate change might take place much more quickly than previously thought, *The Day After Tomorrow*'s fifteen minutes from heat wave to flash freeze was pushing the envelope, to put it mildly.

Even though *The Day After Tomorrow* puts climate change on *extreme* fast-forward, some parts of the movie have no grounding in present-day science at all. For example, the flash freeze in the movie is attributed in part to a downdraft from the upper troposphere. While tropospheric warming has been coupled with stratospheric cooling, no known science yet supports the idea that these two layers of the atmosphere might interact so violently and swiftly. The real story of coming climate change can't be played out in two hours. After a few minutes, watching ice melt one prosaic drop at a time would not sell many tickets at the box office. The real story is inordinately complex, not sound-bite sized.

The New Yorker's reviewer, Anthony Lane, gave *The Day After Tomorrow* an emphatic two thumbs down: "Even by the standards of disaster movies, 'The Day After Tomorrow' is irretrievably poor: a shambles of dud writing and dramatic inconsequence which left me determined to double my consumption of fossil fuels" (Lane 2004, 103). Lane mocked the self-seriousness of the movie. Emmerich's *Godzilla*, by contrast, "came with no health caution about overfeeding our pet iguanas," he wrote, "but 'Day After Tomorrow' is so puffed up with ecological pride that it can hardly move. Worse still, in some quarters, it is being taken seriously. . . . Hence the grim, puritanical deal that is struck by this film: having offered us the undoubted pleasure of watching the Empire State Building turn into the world's tallest Popsicle, it makes us pay for that pleasure by lecturing us on what irresponsible citizens we have been. I can just about take this from politicians, but not, I fear, from the man who directed 'Stargate' " (p. 103).

San Francisco Chronicle science writer Keay Davidson dissected the errant atmospheric physics of the film with help from several climatologists. The speed with which Earth freezes over is "scientifically very unrealistic," said Jarvis Moyers, director of the atmospheric sciences division at the U.S. National Science Foundation in Washington, D.C.

> Consider the scene in which New York City skyscrapers freeze from the top down, as a weird kind of thermal "hurricane" drives super-cold air from the upper atmosphere to the ground. Sounds exciting, no? But there's a big problem: In the real world, a descending parcel of air becomes denser and warmer. This is an elementary principle of air pressure physics, firmly established in the seventeenth century and now taught in high school science classes. It's why the Santa Ana winds of Southern California are so

warm: They grow denser as they rush down mountainsides and warm about 5 degrees for every 1,000 feet of descent. The same thing happens when you inflate a tire: It grows warmer as air molecules squeeze tighter together. (Touch the tire, you'll see.) (Davidson 2004, E-1)

Davidson had ignored (or was unacquainted with) cold-air advection that sometimes conveys a cold air mass to the surface in downdrafts from supercell thunderstorms in summer or severe snowstorms in winter. Such advection is uncommon in San Francisco but well known to weather forecasters and other residents of the U.S. Midwest.

As for the flash freezing of oceans depicted in the movie, Suzanne Moser, who studies climate change at the National Center for Atmospheric Research, said: "It's basic physics—physics doesn't allow this (rapid freezing) because of the lag time you need to move heat through a large system, one that is the size of the Earth. If you stick a glass of water in the freezer, it takes—what?—20 or 30 minutes to turn into ice. Now imagine trying to freeze an entire ocean! Even just to 'lock up' some ocean water in the form of ice caps at the poles would take centuries" (Davidson 2004, E-1).

On the whole, *The Day After Tomorrow* offers "an absurd sort of scenario," said Jan Null, a well-known meteorological consultant in the Bay Area, who accompanied Davidson to a showing of the film (Davidson 2004, E-1). Null cited the film's scientific flaws, such as the giant thermal "hurricanes" that freeze everything beneath them. No such meteorological phenomena exist, nor is there any evidence that they ever have existed, he said. Real-life hurricanes, like those that endanger the Gulf of Mexico and U.S. East Coast, are fueled by the abundant energy of warm, humid, rising air in the tropics. By contrast, the movie's super-cold hurricanes are supposedly fueled by cold polar air, yet they're far bigger than real-life hurricanes. The very notion of continent-size "cold" hurricanes is inane because cold air by definition contains much less energy than warm air (p. E-1).

Scientific or not, *The Day After Tomorrow* opened in the U.S. Midwest during a weekend studded with wild weather, including deadly tornadoes. "If last week's storms didn't get your attention, perhaps the doomsday movie 'The Day After Tomorrow' has. Climate change—sometimes referred to as global warming—is back on the front burner," wrote Eli Kintisch of the *St. Louis Post-Dispatch* (Kintisch 2004, B-1). Replied the skeptics: It's the Midwest, and it's late spring. Weather happens.

Although the filmmakers don't acknowledge it, they're also indebted to the folklore of antievolutionists and creationists, wrote Davidson. In one scene in *The Day After Tomorrow*, young people look at a museum display of a woolly mammoth that supposedly froze instantly many millennia ago, "with its food still in its mouth." The insinuation is that there's paleontological evidence for planetary quick freezes like the one depicted in the film. In fact, said Davidson, the wooly mammoth tale is a variation on a story that has kicked around antievolutionist circles since the nineteenth century. Early creationists argued that such creatures were victims of a sudden global "catastrophe" ordained by God, akin to Noah's flood. Creationists prefer to explain biological change in catastrophic rather than slow, evolutionary terms because the former better jibes with the catastrophe-unleashing God of the Bible (Davidson 2004, E-1).

As a disaster flick, *The Day After Tomorrow* does not rank with the greats, wrote Semple. He thought that "its dialogue is overwrought, its symbolism sophomoric, its subplots annoyingly irrelevant and its relationship to scientific reality tenuous at best" (Semple 2004). The special effects were "terrific," however, and Semple asserted that the film's political timing could not have been better: "Scientists, environmentalists and a few lonely politicians have been trying without great success to get the public and the Bush administration to take global warming seriously, and to inject the issue into a presidential campaign that so far seems determined to ignore it" (Semple 2004).

The political potential of the $125-million film did provoke three weeks' worth of debate regarding a long-range threat to the well-being of the Earth—one reason, according to Semple, that the American Museum of Natural History in Manhattan was happy to present the film's New York City premiere in a theater not far from its Hall of Biodiversity, "biodiversity being one of global warming's most likely victims" (Semple 2004). The liberal political action group MoveOn.org hosted a rally a few hours before *The Day After Tomorrow*'s New York premiere, surrounded by fake snow. In a church a block away, 500 boisterous MoveOn.org members gathered for a global warming briefing and pep rally. Al Gore, according to one critic, "played both movie critic and science lecturer" (Jensen 2004). He called the movie "extremely enjoyable and exciting—beyond the message." Before launching into his humor-studded slide show about carbon dioxide and dwindling glaciers, the former vice president, who wrote the 1992 environmental tome *Earth in the Balance*, praised "the honest fiction of this movie" (Jensen 2004). If one *really* wanted fiction, averred Gore, try the Bush administration's line on global warming.

Some boosters of the film confused cinematic fiction with reality. Laurie David, a trustee of the Natural Resources Defense Council, told the crowd at the museum, "We all know one disaster film is worth 1,000 environmental speeches." David expected the movie to be "the tipping point" in the debate over global warming, as she added, "I never thought I'd be uttering these words, but thank you, Fox" (Jensen 2004).

The politics of the matter were another reason why so many people on both sides of the issue were going to so much trouble to sermonize over what was, in Semple's words, "just a mindless summer blockbuster" (Semple 2004). The Worldwatch Institute, for example, capitalized on the film's publicity, urging its audience to park their automobiles and ride bicycles to theaters. Senator John McCain expressed hope that the film might help win a few votes for a bill he was cosponsoring with Senator Joseph Lieberman that would slowly begin to reduce industrial carbon dioxide emissions in the United States.

Back in the real world, a tornado larger than any depicted in the movie took Hallam, Nebraska, seventy-five miles southwest of Omaha, off the map a week and a half before the film opened. A week later, hurricane-force winds and tennis-ball sized hail knocked out power to half a million homes and businesses in the Dallas-Fort Worth area, including some theaters where the movie was scheduled to open.

"If you are expecting this film to educate you on global warming, forget about it," said Michael Oppenheimer, an adviser to Environmental

Defense and a Princeton University expert on the subject. On the other hand, he added, the movie may well have caused some people to start paying attention to the realities of the issue, which "are plenty bad enough" (Semple 2004). Oppenheimer's calculations suggest that unless the use of fossil fuels is reduced, every beach in the New York City metropolitan area would be lost to rising seas by the end of the twenty-first century. After that, his forecasts anticipate that substantial melting of ice sheets in Greenland and Antarctica could put much of lower Manhattan under water.

"The core question is whether the climatic disruption is a serious issue or not," said George M. Woodwell, director of the Woods Hole Research Center and a longtime scientific explorer of global warming. "The movie says it's a very serious issue. The movie is of course entertainment. It's science fiction in the sense it's taking an issue and carrying it through to a conclusion that is improbable if not totally false. On the other hand, climate disruption is very serious, and it's not being addressed by our government" ("Storm Warning" 2004, N-17). Very few Hollywood disaster films are reviewed in scientific journals, but *The Day After Tomorrow* received some ink (and a qualified endorsement) in the British scientific journal *Nature*: "While unfettered anthropogenic climate change certainly will not turn out exactly like *The Day After Tomorrow*, it should still be a show worth watching" (Allen 2004, 348).

"The good news is that the disaster in the film is greatly exaggerated. The bad news is that the risk of global climate change is truly severe," said James Gustave Speth, dean of the Yale University School of Forestry and Environmental Studies and author of *Red Sky at Morning: America and the Crisis of the Global Environment*, which was being published as *The Day After Tomrrow* was showing (Jurgensen 2004, D-1). "The kind of changes expected in the future are really scary enough that you don't need to go beyond them to these drastic scenarios," said Julia Verville, a climate expert with the Union of Concerned Scientists, which had endorsed the film as an environmental wake-up call (p. D-1).

"Some of this may be due to certain natural variability, but frankly, I don't think there's a serious question anymore whether we're seeing global warming or not. This is happening," Speth said. The consequences may not be blockbuster fodder, he said, but that shouldn't be a requisite for preventive action. "If all of the maple trees disappear out of Connecticut, if all of the coral disappears out of the ocean, if sea level inundates the coastal marshes of Connecticut and we get severe weather events like the hurricane of 1938, these are not unrealistic possibilities at all," Speth said. "It really is an extraordinary sacrifice: the climate future of our country for short-term politics" (Jurgensen 2004, D-1).

Regardless of its scientific merits, *The Day After Tomorrow* certainly put global warming back on the political agenda, at least for a few weeks, much more effectively than many earnest scientific warnings or any number of debates over the efficacy of the Kyoto Protocol. As Geoff Kitney wrote in the *Australian Financial Review:* "They [the British] need only to read the official reports to Prime Minister Tony Blair from his chief scientist, David King. King offers only a slightly less alarming scenario for the impact of global warming based on science than Emmerich does based on trying to scare the pants off his audiences" (Kitney 2004, 30). In advice to Blair before the movie was released, King had asserted (following Hans Blix, former Iraq weapons inspector for the United Nations) that climate change is a bigger global threat than terrorism. He said that the latest data from a wide range of research indicates that the Earth is warming much more rapidly than previously recognized, that this is substantially the result of human activity, and that urgent global political action is required to deal with the threat it poses to life on Earth. Blair publicly supported King's view that global warming is the most serious long-term issue facing the world.

King argued that if fossil fuel emissions are not curtailed significantly, nearly all of the world (except Antarctica) could become uninhabitable by the end of the twenty-first century. "That is," he commented in the *Australian Financial Review*, "children born today may live to see a catastrophic change to life on Earth caused by climate change" (Kitney 2004, 30). King based his forecast on the fact that carbon dioxide levels in the atmosphere by 2100 could be higher than at any time since the days of the dinosaurs. The last time greenhouse gas levels reached the level expected by the end of the twenty-first century, they played a role in a worldwide mass extinction (King 2004, 176–177).

Climate change became a salient political issue across Europe while the federal government of the United States did its best to fixate public attention on the perils of terrorists. In France, for example, after an estimated 15,000 people died of heat-related causes during the August 2003 European heat wave, President Jacques Chirac proposed a constitutional change that would give environmental issues as much weight as human and social rights.

The Day After Tomorrow brought media attention to other global-warming initiatives. For example, the day the movie was released, the *Annapolis* (Maryland) *Capital* featured on its front page a review of the disaster film along with a description of a less spectacular but more scientifically accurate local documentary, *We Are All Smith Islanders*, that had been created by the Montgomery County—based Chesapeake Climate Action Network. This film described problems associated with global warming and sea-level rise on the Chesapeake Bay. *We Are All Smith Islanders* argues that sea-level rise, brought about in part by warmer global temperatures that have made glaciers melt, is causing erosion throughout the bay's watershed (Unger 2004, A-1).

"The entire existence of Smith Island is in jeopardy from sea level rise," said Mike Tidwell, the Chesapeake Climate Action Network's director. "It's wreaking havoc in terms of erosion along every inch of the Chesapeake Bay" (Unger 2004, A-1). The average temperature at the mouth of the York River in the lower bay for the months of December, January, and February has increased 3 degrees F since 1957, according to Bob Wood, acting branch chief of the Oxford Lab of the National Oceanic and Atmospheric Administration. Near Solomon Island, sea level has risen an average of about 3 millimeters per year since 1937, Wood said, as the result of a combination of subsidence and rising sea level (p. A-1).

At about the time *The Day After Tomorrow* was released, a poll conducted by the Yale University School of Forestry and Environmental Studies indicated that 70 percent of United States citizens believed global warming was a very serious or somewhat serious problem; 20 percent said that global warming is not a serious issue ("Yale University" 2004). "The results couldn't be clearer," said Speth. "People have serious concerns about global warming because they believe the scientific data show there is a problem. Americans of all stripes recognize that unless we act now, our world will grow hotter, sea levels will rise, and the Earth could suffer increasingly severe droughts, floods, windstorms, and wildfires. We don't need Hollywood to exaggerate the issue," asserted Speth. "The American public understands that the buildup of greenhouse gases is a very real threat. It is very scary. And people want it dealt with—now" ("Yale University" 2004).

Climate-change skeptics had a heyday with the film's scenes of midsummer snow and heat waves alternating nearly instantly with freakish freezes intense enough to send waves of United States citizens streaming southward into Mexico. "Is this what it has finally come down to?" asked James M. Taylor in the *Boston Globe*. "Rebuffed by science and ignored by the public, global warming alarmists are desperate enough for political relevance to trumpet second-rate Hollywood sensationalism as a 'teachable moment' for the complex science of climate change" (Taylor 2004, A-11).

Veteran climate-change skeptic Patrick J. Michaels used the movie for political mileage, opining in *USA Today* (as well as a number of other newspapers): "As a scientist, I bristle when lies dressed up as 'science' are used to influence political discourse. The latest example is the global-warming disaster flick, *The Day After Tomorrow*" (Michaels 2004, 21-A). Michaels, a professor of environmental science at the University of Virginia, called the movie "off the wall" and "absurd" ("Storm Warning" 2004, N-17). Michaels fumed: "This film is propaganda designed to shift the policy of this nation on climate change. At least that's what I take from producer Mark Gordon's comment that 'part of the reason we made this movie' was to 'raise consciousness about the environment' " (p. 21-A).

"The stratosphere will become the troposphere," wrote Michaels, "when all three laws of thermodynamics are repealed. . . . Hurricanes can't hit Belfast because the intervening island of Ireland would destroy them" (Michaels 2004, 21-A). As for the rising numbers of tornadoes in the U.S. Midwest (more were reported in May 2003 than any other month on record), Michaels believes this has more to do with rising coverage by radar than actual numbers of storms. The tornado "outbreak" is the result of sensitive Doppler radar, according to Michaels (p. 21-A). Bjorn Lomborg, a self-described environmentalist who has become famous an a climate-change skeptic, added: "It is safe to say that global warming will not lead to the onset of a new ice age. . . . It is highly unlikely that global warming will lead to a widespread collapse" of the Gulf Stream (Lomborg 2004). Lomborg, author of *The Skeptical Environmentalist* (Cambridge University Press, 2001), asserted, "The only way to produce an ocean circulation without the Gulf Stream would be to turn off the wind system or stop the Earth's rotation, or both" (Lomborg 2004).

MICHAEL CRICHTON'S *STATE OF FEAR*

Into the weather wars, during 2004, Michael Crichton, seller of 100 million books to date (including *Jurassic Park*), threw *State of Fear*, a 603-page novel published by HarperCollins with a fourteen-page bibliography and a five-page "factual" antiwarming screed. (With "facts" like these, who needs fiction?) *State of Fear* creates a world in which environmentalists fake global warming and murder skeptics to keep the green stuff rolling in. Crichton's fake take puts a new angle on the term "Green Party." Two million copies of this fantasy hit bookstores just before Christmas. Speaking of the green stuff, Crichton's climate-change fable began the year 2005 listed as number two in sales worldwide by Amazon.com.

Crichton's greenies are unnaturally powerful people who are capable of provoking earthquakes, underwater landslides, and a tsunami to cow the gullible into believing that global warming is a threat. And what do earthquakes have to do with global warming? Ask Crichton. Sound science, this most certainly isn't. As described by Michiko Kakutani in the *New York Times*, Crichton's villains are "tree-hugging environmentalists, believers in global warming, proponents of the Kyoto Protocol. Their surveillance operatives drive politically correct, hybrid

Priuses; their hit men use an exotic, poisonous Australian octopus as their weapon of choice. Their unwitting (and sometimes, witting) allies are—natch!—the liberal media, trial lawyers, Hollywood celebrities, mainstream environmental groups (like the Sierra Club and the Audubon Society) and other blue-state apparatchiks." This "ham-handed" novel, Kakutani wrote, "reads like a shrill, preposterous right-wing answer [to the] shrill, preposterous but campily entertaining global warming disaster movie *The Day After Tomorrow*." (Kakutani 2004, E-1).

CHINA: GREENHOUSE-GAS WILD CARD

Right-wing politicians in the United States have opposed the Kyoto Protocol because it exempts developing countries (the two largest being

A cement plant hovers in haze behind a gas station in Shuo Zhou City, Shanxi, China, February 2004. © Peter Essick/Aurora/Getty Images.

China and India) from required reductions in greenhouse gas emissions. The exemption is a measure of economic justice; during 2002, for example, China's per capita greenhouse gas emissions were roughly one-eighth those in the United States. India's were much less.

Regardless of the political situation, China is the wildest card in the world greenhouse deck. On one hand, the world's most populous country is streamlining energy efficiency and experimenting with new fuel sources. On the other, China is undergoing an industrial revolution with a population base of more than 1.3 billion, consuming rapidly increasing amounts of coal and oil even as its economy becomes more efficient.

As China industrializes, its planners have tried to deemphasize the use of coal, but reality has intervened. China has plentiful coal reserves, and most of its new electrical plants are coal-fired, adding measurably to the atmosphere's load of greenhouse gases. In terms of energy generation, coal (at $3 per million British thermal units [BTUs]) was less than half as expensive as oil (at $7) or natural gas (at $8) in 2004. China's coal production increased almost 200 million tons, to 1.9 billion tons, between 2003 and 2004 (Barta and Smith 2004, A-1). China and India account for roughly two-thirds of global demand for coal.

David G. Streets and colleagues estimated that China's carbon dioxide emissions fell 7.3 percent between 1996 (the peak year) and 2000, while its methane emissions fell 2.2 percent between 1997 (the peak year) and 2000, "because China undertook a radical reform of its coal and energy industries" (Streets et al. 2001, 1835). In addition, China's economy suffered during the 1997–1998 Asian recession; many factories closed down or curtailed production because of economic restructuring, causing coal production and consumption to decline.

China later revised upward its estimates of coal consumption for 1999, wiping out half the previously reported reductions ("Research Casts Doubt" 2001, A-16). The *Washington Post* reported, "Other research points to a serious under-reporting of China's consumption of oil, another major pollutant" (p. A-16). Given its economic expansion, China could surpass the United States as the world's leading producer of greenhouse gas emissions within three decades.

China continues to rely on coal for 75 percent of its energy, spewing out some 19 million tons of sulfur dioxide a year (the United States produces 11 million tons per year), and contributing mightily to acid rain (Becker 2004, 80). Coal consumption in China increased by an estimated 10 percent a year between 2000 and 2003. Chinese electricity generation, the main use of coal, jumped 16 percent during the first eight months of 2003 (Bradsher 2003, 1). Many Chinese homes that once used only light bulbs have acquired several appliances, including air conditioners, in recent years.

At the same time, China's fleet of motor vehicles was expanding rapidly. China's consumption of oil also increased from roughly 2.2 million barrels a day in 1988 to 5.2 barrels a day in 2003, or roughly 150 percent in fifteen years (an average of 10 percent a year). The International Energy Agency issued figures from its office in Paris indicating that increases in Chinese greenhouse gas emissions between 2000 and 2030 "will nearly equal the increase from the entire industrialized world" (Bradsher 2003, 1). During the mid-1990s, people in China owned a mere handful of private cars. Private automobile ownership grew by 26 percent between 1996 and 2000, and by 69 percent in 2003 alone (English 2004, 1).

General Motors has forecast that China will account for 18 percent of global growth in automobile sales between 2002 and 2012 (Bradsher 2003, 1). During the 1990s, motor vehicle sales in the Chinese countryside rose from about 40,000 to almost 500,000 per year (Leggett 2001, A-19). Shanghai Automotive Industrial Corp. is planning to license General Motors technology to build a basic pickup truck for China's farmers. The new vehicle, to be called the Combo, will be produced in a nonprofit government car factory. This is one of General Motors' efforts to tap an auto market of "one billion consumers and a fast-growing network of national highways" (p. A-19).

Nobuhiro Horii, who works with the Institute of Developing Economies in Japan, examined how China's Hunan province handled government orders to close coal mines. "He concluded," according to the *Washington Post*, "that local officials told Beijing they had shut the mines, when in fact they kept them open. Interviews with officials in other parts of China led Horii to determine this to be a nationwide problem" ("Research Casts Doubt" 2001, A-16). Horii added that it usually takes about a decade to increase energy efficiency. China's claims that it was making inroads into carbon dioxide production in two years, or even four, were not credible, he asserted. "This is just not possible," Horii added. "Yes, China is increasing energy efficiency, but they are doing it slowly, like everyone else" (p. A-16). A report by the U.S. Embassy in Beijing called the statistical claims of Chinese greenhouse gas reductions "greatly exaggerated," saying they fell "outside the realm of experience of any other country in modern times." The report concluded that China's greenhouse gas emissions "have dropped little, if at all" (p. A-16).

SKEPTICS EMBRACE INLAND COOLING IN ANTARCTICA

In an apparent contradiction of temperature trends across most of the world, some areas of interior Antarctica have cooled steadily for more than two decades. A study led by Peter Doran of the University of Illinois at Chicago found that temperatures in the Dry Valleys near McMurdo Sound in eastern Antarctica have declined at a rate of 1.2 degrees F per decade since 1986. Similar trends have been observed across the continent's interior since 1978. The apparent cooling of inland Antarctica has been used by climate skeptics to refute global warming as an idea, much to the consternation of scientists involved in research.

Doran stressed that although scientists could not explain the falling temperatures, his research "does not change the fact that the planet has warmed up on the whole. The findings simply point out that Antarctica is not responding as expected" (Gugliotta 2002, A-2). Doran also warned that "you don't want to overstate the effects" of the cooling trend, because any rise in sea level caused by global warming this century is expected to come from thermal expansion of existing oceans and not from any theoretical melting of the southern ice cap (p. A-2).

In a paper published by the journal *Nature*'s online edition, Doran and other members of the National Science Foundation's Long-Term Ecological Research Team presented data gathered during years of research in the McMurdo Dry Valleys, a snow-free mountainous desert of chill, arid soils, bleak bedrock outcroppings, and ice-covered lakes that is home to many microscopic invertebrates, mostly nematodes (Gugliotta 2002, A-2).

This fragile ecosystem requires four to six weeks of above-freezing temperatures during the southern summer, Doran said. This period of relative warmth causes melt-water from hillside glaciers to cascade downward in seasonal arroyos that feed life in local lakes. Researchers have found that temperatures had been dropping, not rising, since 1986, with the most pronounced declines in summer and autumn. Glacial ice has not been melting, and so the streams were not flowing, lakes were shrinking, and microorganisms were disappearing (Doran et al. 2002, 517).

Doran said that his team next studied data collected since 1966 from permanent installations throughout the Antarctic. Previous studies had shown overall warming, but the researchers found that these calculations relied disproportionately on readings from the Antarctic Peninsula. When the researchers corrected for this distortion, they found that Antarctica as a whole had become considerably colder. "Temperatures were rising between 1966 and 1978," Doran said, but then they started to fall and have continued falling ever since (Doran et al. 2002, 517). "Our spatial analysis of Antarctic meteorological data demonstrates a net cooling on the Antarctic continent between 1966 and 2000, particularly during summer and autumn. The McMurdo Dry Valleys have cooled by 0.7 degrees C per decade between 1986 and 2000, with similar pronounced seasonal trends. Summer cooling is particularly important to Antarctic terrestrial ecosystems that are poised on the interface of ice and water. . . . Continental Antarctic cooling, especially the seasonality of cooling, poses challenges to models of climate and ecosystem change" (p. 517).

Doran and colleagues did not venture speculation on causes of the temperature decline. They do know that temperatures in the Dry Valleys rise when the wind blows and clouds cover the sky. Doran explained that as winds roll downhill off the Antarctic plateau into the Dry Valleys, the air compresses and heats up as a result, an effect similar to the Chinook winds of the western United States or the dry, warm Santa Ana winds of southern California (Gugliotta 2002, A-2). At the same time, relatively warm summer winds gathering speed over the ocean bring warmer air in from the coast to promote the thaw, he said. Wind generally brings clouds that add to the warming, Doran added (p. A-2). Recently, however, "We're getting a decrease in winds from both directions," Doran said, and, perhaps as a consequence, temperatures in the Dry Valleys are dropping. "It's clearly connected to the winds, but what's controlling the decrease in the winds is not clear" (p. A-2).

Once Doran and colleagues' results were made public, several newspaper reports rushed to simplify them into a worldwide cooling trend, taking this news as a proxy to refute global warming. The authors of the studies expressed caution. One of the scientists involved in studies indicating that the Ross Ice Shelf was thickening, Slawek Tulaczyk of the University of California at Santa Cruz, said that press misinterpretations left him increasingly frustrated by sometimes-careless media coverage of the global warming issue.

When Tulaczyk and Ian Joughin of the Jet Propulsion Laboratory in Pasadena reported in *Science* that the movement of the glacial ice streams in the Ross Ice Shelf appeared to be slowing and allowing the ice to thicken (see Chapter 11, "Melting Antarctic Ice"), a headline over an editorial in the *San Diego Union-Tribune* minced no words: "Scientific Findings Run Counter to Theory of Global Warming." The editorial sarcastically asked: "Oh dear. What will the doomsayers say now? How will they explain away yet two more scientific studies that clearly contradict the global warming orthodoxy" (Davidson 2002, A-8)? A headline in the *National Post*, a Canadian newspaper, declared: "Antarctic Ice Sheet has Stopped Melting, Study Finds" (p. A-8). "Is Another Ice-Age On the Way?" asked an editorial headline in the *Rocky Mountain News* (p. A-8). Analyzing these reports in the *San Francisco Chronicle*, Keay Davidson commented, "Some media mistakenly equated the phenomenon studied by Joughin and Tulaczyk—a change in ice flow rates—with ice melting rates. The mistake contributed to the erroneous belief that the studies constituted, as it were, scientific 'tests' of the global warming theory" (p. A-8).

Contrary to some news reports, "the ice-sheet growth that we have documented in our study area has absolutely nothing to do with any recent climate trends," Tulaczyk said, emphasizing those words in an e-mail to the *San Francisco Chronicle* (Davidson 2002, A-8). The thickening of Antarctic ice in certain regions—especially Ice Stream C of the Whillans Ice Stream, adjacent to the Ross Ice Shelf—results from the complex internal dynamics of the ice itself. These particular ice-flow changes were unrelated to global warming caused by combustion of fossil fuels; such changes occurred for many millennia before the industrial revolution boosted atmospheric levels of heat-trapping gases. The area with the greatest ice thickening is on an ice stream that stopped flowing about 150 years ago (p. A-8).

"I keep repeating to journalists that climate science is much like economics. Both deal with complex systems," Tulaczyk observed. "Just as a single stock going up or down cannot be interpreted as a reliable indicator of economic recovery or collapse, we have to accept the occurrence of contradictory trends in the global climate" (Davidson 2002, A-8). Contrary to some reports attributed to his research, "Global warming is real and happening right now," asserted Doran. He said that the cooling trend in Antarctica appears to be a surprising, regional exception to the overall planetary warming (p. A-8). Doran emphasized: "Our paper does not change the global [temperature] average in any significant way.... Although we have said that more area of the continent is cooling than warming, one just has to look at the paper itself ... to see that it is a close call" (p. A-8). "Our analysis suggests that about two-thirds of the main continent has been cooling in the last 35 years," Doran said. "But there is one-third of the continent that has been warming if you remove the [Antarctic] Peninsula. And with the Peninsula included, it shrinks to 58 percent cooling" (p. A-8). Doran bluntly advised the public: "If you want the facts, you have to go to the original scientific peer-reviewed literature, and avoid the broken-telephone effect of the popular press" (p. A-8).

HOW WARM WAS THE MEDIEVAL WARM PERIOD?

During 2003, a media stir was created by a study promoting the point of view that temperatures had been warmer during the "medieval warm period" from 900 to 1300 CE than during the twentieth century. Willie Soon, a physicist and astronomer, wrote the study with four coauthors: Sallie Baliunas; Sherwood Idso and his son, Craig Idso, who are the former and current presidents of the Center for the Study of Carbon Dioxide and Global Change; and David Legates, a climate researcher at the University of Delaware.

This study was promoted as a product of Harvard University (two of the authors had affiliations with Harvard), but most of it was supported by nonprofit groups with ties to the oil industry. The paper provoked a global storm of e-mail among scientists, some of whom proposed a boycott of the journal that published the study. "Energy interests paid for the study and help finance the groups promoting it. The study illustrates a strategy adopted in the late 1980s to attack the credibility of climate science," said John Topping, president of the Climate Institute. "They saw early on that what they had to do was keep the science at issue," said Topping, a former Republican congressional staff member who founded the institute in 1986 (Nesmith [May 30] 2003).

The study, "Reconstructing Climatic and Environmental Changes of the Past 1,000 Years: A Reappraisal," which purported to be an analysis of data from more than 200 other studies, was underwritten by the American Petroleum Institute, an advocacy association for the world's largest oil companies. Two of the five authors received support from the ExxonMobil Foundation. Two others, affiliated with the Harvard-Smithsonian Center for Astrophysics, also were listed as "senior scientists" with a Washington-based organization supported by several right-wing foundations and ExxonMobil. One of these organizations, the George T. Marshall Institute, is headed by William O'Keefe, a former executive of the American Petroleum Institute. O'Keefe also is a past president of the Global Climate Coalition, a defunct group created by oil and coal interests to lobby against U.S. participation in treaties such as the Kyoto Protocol (Nesmith May 30, 2003).

Raymond F. Bradley, Malcolm K. Hughes, and Henry F. Diaz, writing in *Science*, asserted that while comparisons of the medieval period with the present are difficult because records were sparse (especially on a worldwide scale), "Temperatures from 1000 to 1200 CE (or 1000 to 1100 CE) were ... 0.03 degrees C cooler than the period from 1901 to 1970 CE. The latter period was on average about 0.35 degrees C cooler than the last thirty years of the twentieth century" (Bradley, Hughes, and Diaz 2003, 405). The authors also asserted that the period 1100 CE to 1260 CE was characterized by high levels of explosive volcanism, which, in our time, have been associated with warmer-than-usual winters in northern Europe and northwestern Russia that may have helped support Viking colonization of Iceland and Greenland. Most of the weather records of the time also come from Europe, which might make the data a poor proxy for worldwide averages.

DEBATE OVER SATELLITE DATA

Climate-change skeptics for nearly two decades have argued that global temperatures as measured by satellites show little or no warming, contrary to most ground-level observations. The skeptics argue that the satellite records are more comprehensive than the ground-level readings, especially over the two-thirds of the Earth that is covered by oceans, where surface measurements are sparse.

By the end of the 1990s, however, older satellite records had been refined and corrected, and the skeptics' case was falling apart. Analysis of satellite data collected between 1979 and 1999 from the lowest few miles of the atmosphere indicated a global temperature rise of about one-third of a degree F between 1979 and 1999. The results are at odds with previous analyses that show virtually no warming in the satellite record over the same period. The findings were published by the journal *Science* at its Science Express Web site (www.sciencexpress.org) on May 1, 2003 ("New Look" 2003).

The scientists who compiled the study included Tom Wigley, Gerald Meehl, Caspar Ammann, Julie Arblaster, Thomas Bettge, and Warren

Washington, all from the National Center for Atmospheric Research. The lead author of the study was Ben Santer of Lawrence Livermore National Laboratory. "It's undeniable that the agreement with both global climate models and surface data is better for the new analysis than for the old one," said Wigley ("New Look" 2003). Instruments aboard twelve U.S. satellites provided temperature records for the lower stratosphere. Each sensor intercepted microwaves emitted by various parts of the atmosphere, with emissions increasing as temperatures rise. These data were used to infer temperatures ("New Look" 2003).

Since the 1990s, climate-change skeptics have used satellite data to argue the absence of a warming signal. Their figures indicate no warming at higher levels of the atmosphere, contrary to distinct warming at Earth's surface. A 2000 report from the National Research Council concluded that both trends might be correct—in other words, the global atmosphere might be warming more quickly near the ground than higher. "The real issue is the trend in the satellite data from 1979 onward," said Wigley. "If the original analysis of the satellite data were right, then something must be missing in the models. With the new data set, the agreement with the models is improved, and the agreement with the surface data is quite good" ("New Look" 2003).

As time has passed, the skeptics' case that warming is not supported by satellite data has continued to deteriorate. In 2003, Konstantin Y. Vinnikov and Norman C. Grody reported an analysis of global tropospheric temperatures from 1978 to 2002, using passive microwave sounding data from the National Oceanic and Atmospheric Administration's series of polar orbiters and the Earth Observing System Aqua satellite. Their analysis showed a trend of plus 0.22 to 0.26 degrees C per decade, "consistent with the global warming trend derived from surface meteorological stations" (Vinnikov and Grody 2003, 269).

REFERENCES: PART I. GLOBAL WARMING SCIENCE: THE EVOLVING PARADIGM

Abrahamson, Dean Edwin. *The Challenge of Global Warming*. Washington, D.C.: Island Press, 1989.

Abram, Nerilie J., Michael K. Gagan, Malcolm T. McCulloch, John Chappell, and Wahyoe S. Hantoro. "Coral Reef Death during the 1997 Indian Ocean Dipole Linked to Indonesian Wildfires." *Science* 301 (August 15, 2003): 952–955.

"Acid Rain Emissions Halve in 11 Years." Hermes Database (Great Britain), May 20, 2003. (Lexis).

Allen, Myles. "Film: Making Heavy Weather." *Nature* 429 (June 7, 2004): 347–348.

Alley, R. B., J. Marotzke, W. D. Nordhaus, J. T. Overpeck, D. M. Peteet, R. A. Pielke Jr., et al. "Abrupt Climate Change." *Science* 299 (March 28, 2003): 2005–2010.

Alleyne, Richard, and Ben Fenton. "Heatwave Britain—When the Trees Turn Toxic." *Daily Telegraph* (London), May 10, 2004, 3.

Anderson, David M., Jonathan T. Overpeck, and Anil K. Gupta. "Increase in the Asian Southwest Monsoon during the Past Four Centuries." *Science* 297 (July 26, 2002): 596–599.

Anderson, J. G., W. H. Brune, and M. H. Proffitt. "Ozone Destruction by Chlorine Radicals within the Antarctic Vortex: The Spatial and Temporal Evolution of ClO/O_3, Anticorrelation Based on In Situ ER-2 Data." *Journal of Geophysical Research* 94 (1989): 11465–11479.

Anderson, J. W. "The History of Climate Change as a Political Issue." The Weathervane: A Global Forum on Climate Policy Presented by Resources for the Future, August 1999. www.weathervane.rff.org/features/feature 005. html.

Anderson, Theodore L., Robert J. Charlson, Stephen E. Schwartz, Reto Knutti, Olivier Boucher, Henning Rodhe, et al. "Climate Forcing by Aerosols—a Hazy Picture." *Science* 300 (May 16, 2003): 1103–1104.

Andreae, M. O., D. Rosenfeld, P. Artaxo, A. A. Casta, G. P. Frank, K. M. Longo, et al. "Smoking Rain Clouds over the Amazon." *Science* 303 (February 27, 2004): 1337–1442.

Andreae, Meinrat O. "The Dark Side of Aerosols." *Nature* 409 (February 8, 2001): 671–672.

———, Chris D. Jones, and Peter M. Cox. "Strong Present-Day Aerosol Cooling Implies a Hot Future." *Nature* 435 (June 30, 2005): 1187–1190.

Arrhenius, Svante. "On the Influence of Carbonic Acid in the Air Upon the Temperature of the Ground." *The London, Edinburgh, and Dublin Philosophical Magazine and Journal of Science*, 5th ser. (April 1896): 237–276.

"Atmospheric Science: Really High Clouds." *Science* 292 (April 13, 2001):171.

Austin, J., N. Butchart, and K. P. Shine. "Possibility of an Arctic Ozone Hole in a Doubled-CO_2 Climate." *Nature* 360 (November 19, 1992): 221–225.

Ball, Philip. "Climate Change Set to Poke Holes in Ozone: Arctic Clouds Could Make Ozone Depletion Three Times Worse Than Predicted." *Nature* Science Update, March 3, 2004. http://info.nature.com/cgi-bin24/DM/y/eOCBoBfHSKoChoJVVoAY.

Barta, Patrick, and Rebecca Smith. "Global Surge in Use of Coal Alters Energy Equation." *Wall Street Journal*, November 16, 2004, A-1, A-17.

Becker, Jasper. "China's Growing Pains: More Money, More Stuff, More Problems. Any Solutions?" *National Geographic*, March 2004, 68–95.

Bengtsson, Lennart O., and Claus U. Hammer. *Geosphere-Biosphere Interactions and Climate.* Cambridge: Cambridge University Press, 2001.

Bernard, Harold W., Jr. *Global Warming: Signs to Watch For.* Bloomington: Indiana University Press, 1993.

Betsill, Michelle M. "Impacts of Stratospheric Ozone Depletion." In *Handbook of Weather, Climate, and Water: Atmospheric Chemistry, Hydrology, and Societal Impacts*, ed. Thomas D. Potter and Bradley R. Colman, 913–923. Hoboken, NJ: Wiley Interscience, 2003.

"Biomass Burning Boosts Stratospheric Moisture." Environment News Service, February 20, 2002. http://ens-news.com/ens/feb2002/2002L-02-20-09.html.

Black, David E. "The Rains May Be A-comin'." *Science* 297 (July 26, 2002): 528–529.

Bodhaine, Barry, Ellsworth Dutton, and Renee Tatusko. "Assessment of Ultraviolet (UV) Variability in the Alaskan Arctic." Cooperative Institute for Arctic Research, University of Alaska, and NOAA, March 6, 2001. www.cifar.uaf.edu/ario0/bodhaine.html.

Bond, Gerard, Bernd Kromer, Juerg Beer, Raimund Muscheler, Michael N. Evans, William Showers, et al. "Persistent Solar Influence on North Atlantic Climate during the Holocene." *Science* 294 (December 7, 2001): 2130–2136.

Bowen, Gabriel J., David J. Beerling, Paul L. Koch, James C. Zachos, and Thomas Quattlebaum. "A Humid Climate State during the Palaeocene/Eocene Thermal Maximum." *Nature* 432 (November 25, 2004): 495–499.

Bowman, Lee. "Soot Could Be Causing a Lot of Bad Weather." Scripps Howard News Service, September 26, 2002, Lexis.

Bradley, Raymond S., Malcolm K. Hughes, and Henry F. Diaz. "Climate in Medieval Time." *Science* 302 (October 17, 2003): 404–405.

Bradsher, Keith. "China Prospering but Polluting: Dirty Fuels Power Economic Growth." *New York Times* in *International Herald-Tribune*, October 22, 2003, 1.

Brasseur, Guy P., Anne K. Smith, Rashid Khosravi, Theresa Huang, Stacy Walters, Simon Chabrillat, et al. "Natural and Human-Induced Pertubations in the Middle Atmosphere: A Short Tutorial." In *Atmospheric Science across the Stratopause*, ed. David E. Siskind et al. Washington, D.C.: American Geophysical Union, 2000.

Broecker, Wallace S. "Was the Medieval Warm Period Global?" *Science* 291 (February 23, 2001): 1497–1499.

Brook, Edward J. "Tiny Bubbles Tell All." *Science* 310 (November 25, 2005): 1285–1287.

Brown, Paul. "Global Warming: Worse Than We Thought." *World Press Review*, February 1999, 44.

Buencamino, Manuel. "Coming Catastrophe?" *BusinessWorld*, June 14, 2004, 21.

"Bush Says Administration Has Not Changed Stance on Global Warming." *The Frontrunner*, August 27, 2004. (Lexis).

Byers, Stephen, and Olympia Snowe, cochairs, International Climate Change Task Force. *Meeting the Climate Challenge: Recommendations of the International Climate Change Task Force.* London: Institute for Public Policy Research, January 2005.

Calamai, Peter. "Alert over Shrinking Ozone Layer." *Toronto Star*, March 18, 2002, A-8.

Callendar, G. D. "The Artificial Production of Carbon Dioxide and Its Influence on Temperature." *Quarterly Journal of the Royal Meteorological Society* 64 (1938): 223–240.

"Carbon Sinks Cannot Keep Up with Emissions." Environment News Service, May 16, 2002. http://ens-news.com/ens/may2002/2002L-05-16-09.html#anchor3.

Chameides, William L., and Michael Bergin. "Climate Change: Soot Takes Center Stage." *Science* 297 (September 27, 2002): 2214–2215.

Charlson, Robert J., John H. Seinfeld, Athanasios Nenes, Markku Kulmala, Ari Laaksonen, and M. Cristina Facchini. "Reshaping the Theory of Cloud Formation." *Science* 292 (June 15, 2001): 2025–2026.

Christensen, Torben R., Torbjörn Johansson, H. Jonas Åkerman, Mihail Mastepanov, Nils Malmer, Thomas Friborg, et al. "Thawing Sub-Arctic Permafrost: Effects on Vegetation and Methane Emissions." *Geophysical Research Letters* 31 (4) (February 20, 2004).

Christianson, Gale E. *Greenhouse: The 200-Year Story of Global Warming.* New York: Walker and Company, 1999.

Christie, Maureen. *The Ozone Layer: A Philosophy of Science Perspective.* Cambridge: Cambridge University Press, 2001.

Ciborowski, Peter. "Sources, Sinks, Trends, and Opportunities." In *The Challenge of Global Warming,* ed. Edwin Abrahamson, 213–230. Washington, D.C.: Island Press, 1989.

Clark, Peter. U., N. G. Pisias, T. F. Stocker, and A. J. Weaver. "The Role of the Thermohaline Circulation in Abrupt Climate Change." *Nature* 415 (February 21, 2002): 863–868.

Clift, Peter, and Karen Bice. "Earth Science: Baked Alaska." *Nature* 419 (September 12, 2002): 129–130.

Cline, William R. *The Economics of Global Warming.* Washington, D.C.: Institute for International Economics, 1992.

"Clouds, but No Silver Lining." *Guardian* (London), January 24, 2002.

Connor, Steve. "Global Warming Is Twice as Bad as Previously Thought." *Independent* (London), January 27, 2005, 10.

———. "Peat Bog Gases Accelerate Global Warming." *London Independent,* July 8, 2004, 9.

"Contrails Linked to Temperature Changes." Environment News Service, August 8, 2002. http://ens-news.com/ens/aug2002/2002-08-08-09.asp#anchor4.

Cooke, Robert. "Global Warming Is in the Air." *Newsday,* September 20, 2002, A-34.

Cowen, Robert C. "One Large, Overlooked Factor in Global Warming: Tropical Forest Fires." *Christian Science Monitor,* November 7, 2002, 14.

Cox, Peter M., Richard A. Betts, Chris D. Jones, Steven A. Spall, and Ian J. Totterdell. "Acceleration of Global Warming Due to Carbon-Cycle Feedbacks in a Coupled Climate Model." *Nature* 408 (November 9, 2000): 184–187.

Crutzen, Paul J. "The Antarctic Ozone Hole, a Human-Caused Chemical Instability in the Stratosphere: What Should We Learn from It? In *Geosphere-Biosphere Interactions and Climate*, ed. Lennart O. Bengtsson and Claus U. Hammer, 1–11. Cambridge: Cambridge University Press, 2001.

Davidson, Keay. "Film's Tale of Icy Disaster Leaves the Experts Cold." *San Francisco Chronicle*, June 1, 2004, E-1.

———. "Media Goofed on Antarctic Data: Global Warming Interpretation Irks Scientists." *San Francisco Chronicle*, February 4, 2002, A-8.

———. "Study Has New Evidence of Global Warming: Data, Taken 27 Years Apart, Shows Less Heat Escaping Earth Now." *San Francisco Chronicle*, March 15, 2001, A-2.

Davis, Neil. *Permafrost: A Guide to Frozen Ground in Transition*. Fairbanks: University of Alaska Press, 2001.

Dayton, Leigh. " 'Scary' Science Finds Earth Heating up Twice as Fast as Thought." *Australian* (Sydney), January 27, 2005, 3.

del Giorgio, Paul A., and Carlos M. Duarte. "Respiration in the Open Ocean." *Nature* 420 (2002): 379–384.

Diaz, Henry F., and Raymond S. Bradley. "Temperature Variations during the Last Century at High-Elevation Sites." *Climatic Change* 36 (1997): 253–279.

Dickens, Gerald R. "A Methane Trigger for Rapid Warming?" Review of *Methane Hydrates in Quaternary Climate Change: The Clathrate Gun Hypothesis*, by James P. Kennett, Kevin G. Cannariato, Ingrid L. Hendy, and Richard J. Behl. *Science* 299 (2003): 1017.

———. "Global Change: Hydrocarbon-Driven Warming." *Nature* 429 (June 3, 2004): 513–515.

Doran, Peter T., John C. Priscu, W. Berry Lyons, John E. Walsh, Andrew G. Fountain, Diane M. McKnight, et al. "Antarctic Climate Cooling and Terrestrial Ecosystem Response." *Nature* 415 (January 30, 2002): 517–520.

"Dying Trees Release Air Polluting Chemical." Environment News Service, June 26, 2002. http://ens-news.com/ens/jun2002/2002-06-26-09.asp#anchor3.

Easterling, David, et al. "Temperature Range Narrows between Daytime Highs and Nighttime Lows." *Science* (July 18, 1997).

"Eco Bridge: What Can We Do about Global Warming?" No date. www.ecobridge.org/content/g_wdo.htm.

"Editors' Choice: A Summary of Glaciation." *Science* 295 (January 18, 2002): 401.

English, Andrew. "Feeding the Dragon: How Western Car-Makers Are Ignoring Ecological Dangers in Their Rush to Exploit a Wide-Open Market." *Daily Telegraph* (London), October 30, 2004, 1.

Erickson, Jim. "Boulder Team Sees Obstacle to Saving Ozone Layer: 'Rocks' in Arctic Clouds Hold Harmful Chemicals." *Denver Rocky Mountain News*, February 9, 2001, 37-A.

Fahey, D. W., R. S. Gao, K. S. Carslaw, J. Kettleborough, P. J. Popp, M. J. Northway, et al. "The Detection of Large HNO_3-Containing Particles in the Winter Arctic Stratosphere." *Science* 291 (February 9, 2001): 1026–1031.

Falkowski, P., R. J. Scholes, E. Boyle, J. Canadell, D. Canfield, J. Elser, et al. "The Global Carbon Cycle: A Test of Our Knowledge of Earth as a System." *Science* 290 (October 13, 2000): 291–296.

Fang, Jingyun, Anping Chen, Changhui Peng, Shuqing Zhao, and Longjun Ci. "Changes in Forest Biomass Carbon Storage in China Between 1949 and 1998." *Science* 292 (June 22, 2001): 2320–2322.

Farman, J. C., B. G. Gardiner, and J. D. Shanklin. "Large Losses of Total Ozone Reveal Seasonal Clox/NOx Interaction." *Nature* 315 (1985): 207–210.

Feely, Richard A., Christopher L. Sabine, Kitack Lee, Will Berelson, Joanie Kleypas, Victoria J. Fabry, et al. "Impact of Anthropogenic CO_2 on the $CaCO_3$ System in the Oceans." *Science* 305 (July 16, 2004): 362–366.

Ferguson, H. L. "The Changing Atmosphere: Implications for Global Security." In *The Challenge of Global Warming*, ed. Dean Edwin Abrahamson, 48–62. Washington, D.C.: Island Press, 1989.

Fialka, John J. "Soot Storm: A Dirty Discovery over Indian Ocean Sets off a Fight." *Wall Street Journal*, May 6, 2003, A-1, A-6.

Freeman, C., C. D. Evans, and D. T. Monteith. "Export of Organic Carbon from Peat Soils." *Nature* 412 (August 23, 2001): 785.

———, C., N. Fenner, N. J. Ostle, H. Kang, D. J. Dowrick, B. Reynolds, et al. "Export of Dissolved Carbon from Peatlands under Elevated Carbon Dioxide Levels." *Nature* 430 (July 8, 2004): 195–198.

Freeman, James. "Methane Bubbles That Could Sink Ships." *Glasgow (Scotland) Herald*, November 2, 2000, 7.

Giardina, C., and M. Ryan. "Evidence That Decomposition Rates of Organic Carbon in Mineral Soil Do Not Vary with Temperature." *Nature* 404 (2000): 858–861.

Gill, R. A., H. W. Polley, H. B. Johnson, L. J. Anderson, H. Maherali, and R. B. Jackson. "Nonlinear Grassland Responses to Past and Future Atmospheric CO_2." *Nature* 417 (May 16, 2002): 279–282.

Glantz, Michael H. *Climate Affairs: A Primer*. Washington, D.C.: Island Press, 2003.

"Global Warming Could Persist for Centuries." Environment News Service, February 18, 2002. http://ens-news.com/ens/feb2002/2002L-02-18-09.html.

"Global Warming's Sooty Smokescreen Revealed." New Scientist.com, June 3, 2003. www.newscientist.com/news/news.jsp?id=ns99993798.

Goodale, Christine L., and Eric A. Davidson. "Carbon Cycle: Uncertain Sinks in the Shrubs." *Nature* 418 (August 8, 2002): 593–594.

Gordon, Anita, and David Suzuki. *It's A Matter of Survival.* Cambridge, MA: Harvard University Press, 1991.

Goulden, M. L., S. C. Wofsy, J. W. Harden, S. E. Trumbone, P. M. Crill, S. T. Gower, et al. "Sensitivity of Boreal Forest Carbon Dioxide to Soil Thaw." *Science* 279 (January 9, 1998): 214–217.

Graf, Hans-F. "The Complex Interaction of Aerosols and Clouds." *Science* 303 (February 27, 2004): 1309–1311.

Gugliotta, Guy. "In Antarctica, No Warming Trend; Scientists Find Temperatures Have Gotten Colder in Past Two Decades." *Washington Post*, January 14, 2002, A-2.

Haigh, Joanna D. "Climate Variability and the Influence of the Sun." *Science* 294 (December 7, 2001): 2109–2111.

"Half U.S. Climate Warming Due to Land Use Changes." Environment News Service, May 28, 2003. http://ens-news.com/ens/may2003/2003-05-28-01.asp.

Hansen, James, D. Johnson, A. Lacis, S. Lebedeff, P. Lee, D. Rind, et al. "Climate Impact of Increasing Atmospheric Carbon Dioxide." *Science* 213 (August 28, 1981): 957–966.

———, and Larissa Nazarenko. "Soot Climate Forcing via Snow and Ice Albedos." *Proceedings of the National Academy of Sciences* 101 (2) (2004): 423–428.

———, R. Ruedy, A. Lacis, D. Koch, I. Tegen, T. Hall, et al. "Climate Forcings in Goddard Institute for Space Studies SI2000 Simulations." *Journal of Geophysical Research* 107 (D18) (2002): 4347. doi:10.1029/2001JD001143.

Hansen, James E. "Dangerous Anthropogenic Interference: A Discussion of Humanity's Faustian Climate Bargain and the Payments Coming Due." Presented in the Distinguished Public Lecture Series, Department of Physic and Astronomy, University of Iowa, October 26, 2004.

———. "The Greenhouse, the White House, and Our House." Typescript of a speech presented at the International Platform Association, Washington, D.C., August 3, 1989.

Harries, J. E., H. E. Brindley, P. J. Sagoo, and R. J. Bantges. "Increases in Greenhouse Forcing Inferred from the Outgoing Long-Wave Radiation Spectra of the Earth in 1970 and 1997." *Nature* 410 (March 15, 2001): 355–357.

Heilprin, John. "Study Says Black Carbon Emissions in China and India Have Climate Change Effects." Associated Press, September 26, 2002. (Lexis).

Henderson, Mark. "Hot News from 740,000 Years Ago Tells Us to Get Ready for Catastrophic Climate Change." *Times* (London), June 10, 2004, 4.

————. "Positive Winds Keeping Arctic Winters at Bay." *Times* (London), July 6, 2001. (Lexis).

Hinrichs, Kai-Uwe, Laura R. Hmelo, and Sean P. Sylva. "Molecular Fossil Record of Elevated Methane Levels in Late Pleistocene Coastal Waters." *Science* 299 (February 21, 2003): 1214–1217.

Hoerling, Martin P., and Arun Kumar. "The Perfect Ocean for Drought." *Science* 299 (January 31, 2003): 691–694.

Hoffheiser, Chuck. Letter to the editor. *Wall Street Journal*, June 19, 2001, A-23.

Houghton, John T. *Global Warming: The Complete Briefing*. Cambridge: Cambridge University Press, 1997.

Houlder, Vanessa. "Faster Global Warming Predicted: Met Office Research Has 'Mind-Blowing' Implications." *The Financial Times* (London), November 9, 2000, 2.

Hu, Feng Sheng, Darrell Kaufman, Sumiko Yoneji, David Nelson, Aldo Shemesh, Yongsong Huang, et al. "Cyclic Variation and Solar Forcing of Holocene Climate in the Alaskan Subarctic." *Science* 301 (September 26, 2003): 1890–1893.

Human, Katy, and Kim McGuire. "Ozone Decline Stuns Scientists." *Denver Post*, March 2, 2005, A-8.

Ingham, John. "Fears for the Earth. All Down to Us: Hotter by the Minute." *Express* (London), September 2, 2003, 19.

"Inhofe Calls Global Warming Warnings a Hoax." Associated Press, Oklahoma State, and Local Wire, July 29, 2003. (Lexis).

Ivins, Molly. "Ignoring Problem Works—For a While." *Charleston (WV) Gazette*, June 28, 2003, 4-A.

Jackson, Robert B., Jay L. Banner, Esteban G. Jobbagy, William T. Pockman, and Diana H. Wall. "Ecosystem Carbon Loss with Woody Plant Invasion of Grasslands." *Nature* 418 (August 8, 2002): 623–626.

Jacobson, Mark. "Strong Radiative Heating Due to the Mixing State of Black Carbon in Atmospheric Aerosols." *Nature* 409 (February 8, 2001): 695–697.

Jager, J., and H. L. Ferguson. *Climate Change: Science, Impacts, and Policy; Proceedings of the Second World Climate Conference*. Cambridge: Cambridge University Press, 1991.

Jensen, Elizabeth. "Activists Take 'The Day After' for a Spin." *Los Angeles Times*, May 26, 2004. (Lexis).

Jones, C. D., P. M. Cox, R. L. H. Essery, D. L. Roberts, and M. J. Woodage. "Strong Carbon Cycle Feedbacks in a Climate Model with Interactive CO_2 and Sulphate Aerosols." *Geophysical Research Letters* 30(9) (2003), 16,867. doi: 0.1029/2003GL0.

Jucks, K. W., and R. J. Salawitch. "Future Changes in Atmospheric Ozone." In *Atmospheric Science across the Stratosphere*, ed. David E. Siskind, Stephen D. Eckermann, and Michael E. Summers, 241–256.Washington, D.C.: American Geophysical Union, 2000.

Jurgensen, John. "The Weather End Game: The Climate-Change Disaster at the Heart of 'Day After Tomorrow' May be Overplayed, but the Global-Warming Threat Is Real." *Hartford (CT) Courant*, May 27, 2004, D-1.

Kakutani, Michiko. "Beware! Tree-Huggers Plot Evil to Save World." *New York Times*, December 13, 2004, E-1.

Kalnay, Eugenia, and Ming Cai. "Impact of Urbanization and Land-Use Change on Climate." *Nature* 423 (May 29, 2003): 528–531.

Kane, R. L., and D. W. South. "The Likely Roles of Fossil Fuels in the Next 15, 50, and 100 Years, with or without Active Controls on Greenhouse-Gas Emissions." In *Limiting Greenhouse Effects: Controlling Carbon-Dioxide Emissions. Report of the Dahlem Workshop on Limiting the Greenhouse Effect, Berlin, September 9–14, 1990*, ed. G. I. Pearman, 189–227. New York: John Wiley & Sons, 1991.

Keeling, Charles. D., and Timothy. P. Whorf. "Atmospheric CO_2 Records from Sites in the SIO Air Sampling Network." In *Trends: A Compendium of Data on Global Change*. Oak Ridge, TN: Carbon Dioxide Information Analysis Center, Oak Ridge National Laboratory, U.S. Department of Energy, 2004.

Kellogg, William W. "Theory of Climate Transition from Academic Challenge to Global Imperative." In *Greenhouse Glasnost: The Crisis of Global Warming*, ed. Terrell J. Minger, 99. New York: Ecco Press, 1990.

Kennett, James P., Kevin G. Cannariato, Ingrid L. Hendy, and Richard J. Behl. *Methane Hydrates in Quaternary Climate Change: The Clathrate Gun Hypothesis.* Washington, D.C.: American Geophysical Union, 2003.

Kerr, Richard A. "A Perfect Ocean for Four Years of Globe-Girdling Drought." *Science* 299 (January 31, 2003): 636.

———. "A Single Climate Mover for Antarctica." *Science* 296 (May 3, 2002): 825–826.

———. "Stratospheric 'Rocks' May Bode Ill for Ozone." *Science* 291 (February 9, 2001): 962–963.

———. "A Variable Sun Paces Millennial Climate." *Science* 294 (November 16, 2001): 1431–1432.

Keys, David. "Global Warming: Methane Threatens to Repeat Ice-Age Meltdown; 55 Million Years Ago, a Massive Blast of Gas Drove Up Earth's Temperature 7 C and Another Explosion Is in the Cards, Say the Experts." *Independent* (London), June 16, 2001. (Lexis).

King, David A. "Climate Change Science; Adapt, Mitigate, or Ignore?" *Science* 303 (January 9, 2004): 176–177.

Kintisch, Eli. "Floods? Droughts? More Storms? Predictions Vary, but Scientists Agree: Our Climate Will Change; Midwesterners Could End Up Wet, or Dry." *St. Louis Post-Dispatch*, May 30, 2004, B-1.

Kirby, Alex. "Costing the Earth." British Broadcasting Corporation, Radio Four, October 26, 2000. http://news.bbc.co.uk/hi/english/sci/tech/newsid_990000/990391.stm.

Kirk-Davidoff, Daniel, Daniel P. Schrag, and James G. Anderson. "On the Feedback of Stratospheric Clouds on Polar Climate." *Geophysical Research Letters* 29 (11) (2002): 14659–14663.

Kitney, Geoff. "Global Warming Back on the Agenda." *Australian Financial Review*, May 29, 2004, 30.

Knorr, W., I. C. Prentice, J. I. House, and E. A. Holland. "Long-Term Sensitivity of Soil Carbon Turnover to Warming." *Nature* 433 (January 20, 2005): 298–301.

Kondratyev, Kirill, Vladimir F. Krapivin, and Costas A. Varotsos. *Global Carbon Cycle and Climate Change.* Berlin: Springer/Praxis, 2004.

Koren, Ilan, Yoram J. Kaufman, Lorraine A. Remer, and Jose V. Martins. "Measurement of the Effects of Amazon Smoke on the Inhibition of Cloud Formation." *Science* 303 (February 27, 2004): 1342–1345.

Kump, Lee R. "Chill Taken out of the Tropics." *Nature* 413 (October 4, 2001): 470–471.

———. "What Drives Climate?" *Nature* 408 (December 7, 2000): 651–652.

Lacis, Andrew. Personal communication (manuscript critique). Received February 10, 2005.

Lackner, Kalus S. "A Guide to CO_2 Sequestration." *Science* 300 (June 13, 2003): 1677–1678.

Landsea, Christopher. "NOAA: Report on Intensity of Tropical Cyclones." NOAA, Miami, Florida, August 12, 1999. www.aoml.noaa.gov/hrd/tcfaq/tcfaqG.html#G3.

Lane, Anthony. "Cold Comfort: 'The Day After Tomorrow.'" *The New Yorker*, June 7, 2004, 102–103.

Lazaroff, Cat. "Land Use Rivals Greenhouse Gases in Changing Climate." Environment News Service, October 2, 2002. http://ens-news.com/ens/oct2002/2002-10-02-06.asp.

———. "Melting Arctic Permafrost May Accelerate Global Warming." Environment News Service, February 7, 2001. http://ens-news.com/ens/feb2001/2001L-02-07-06.html.

———. "Replacing Grass with Trees May Release Carbon." Environment News Service, August 8, 2002. http://ens-news.com/ens/aug2002/2002-08-08-07.asp.

Lea, David W. "The 100,000-Year Cycle in Tropical SST [Sea-Surface Temperature], Greenhouse Forcing, and Climate Sensitivity." *Journal of Climate* 17 (11) (June 1, 2004): 2170–2179.

Leake, Jonathan. "Fiery Venus Used to Be Our Green Twin." *Sunday Times* (London), December 15, 2002, 11.

Lean, Geoffrey. "Global Warming Will Redraw Map of the World." *Independent* (London), November 7, 2004, 8.

Lean, Judith, and David Rind. "Earth's Response to a Variable Sun." *Science* 292 (April 13, 2001): 234–236.

Leggett, Jeremy, ed. *Global Warming: The Greenpeace Report.* New York: Oxford University Press, 1990, 83–112.

Leggett, Karby. "In Rural China, G[eneral] [Motors] Sees a Frugal but Huge Market: It Bets Tractor Substitute Will Look Pretty Good to Cold, Wet Farmers." *Wall Street Journal*, January 16, 2001, A-19.

Leidig, Michael, and Roya Nikkhah. "The Truth about Global Warming—It's the Sun That's to Blame; The Output of Solar Energy Is Higher Now Than for 1,000 Years, Scientists Calculate." *Sunday Telegraph* (London), July 18, 2004, 5.

Lomborg, Bjorn. "Entertaining Discredited Ideas of a Climatic Catastrophe." *The Australian* (Sydney), May 27, 2004, Lexis.

Lovett, Richard A. "Global Warming: Rain Might Be Leading Carbon Sink Factor." *Science* 296 (June 7, 2002): 1787.

Loya, Wendy M., and Paul Grogan. "Carbon Conundrum on the Tundra." *Nature* 431 (September 23, 2004): 406–407.

Macey, Richard. "Climate Change Link to Clearing." *Morning Herald* (Sydney), June 29, 2004, 2.

Mack, Michelle C., Edward A. G. Schuur, M. Syndonia Bret-Harte, Gaius R. Shaver, and F. Sturt Chapin III. "Ecosystem Carbon Storage in Arctic Tundra Reduced by Long-Term Nutrient Fertilization." *Nature* 432 (September 23, 2004): 440–443.

"Major Temperature Rise Recorded in Arctic This Year: German Scientists." Agence France Presse, August 27, 2004. (Lexis).

Malone, Thomas F., Edward D. Goldberg, and Walter H. Munk. Biographic Memoir for Roger Randall Dougan Revelle, 1909–1991. National Academy of Sciences. No date. www.nap.edu/readingroom/books/biomems/rrevelle.html.

Mann, Michael E., and Philip D. Jones. "Global Surface Temperatures over the Past Two Millennia." *Geophysical Research Letters* 30 (15) (August 2003). doi: 10.1029/2003GL017814. www.ngdc.noaa.gov/paleo/pubs/mann2003b/mann 2003b.html.

McCarthy, Michael. "Countdown to Global Catastrophe." *Independent* (London), January 24, 2005, 1.

McCrea, Steve. "Air Travel: Eco-tourism's Hidden Pollution." *San Diego Earth Times*, August 1996. www.sdearthtimes.com/et0896/et0896s13.html.

McFarling, Usha Lee. "Fear Growing over a Sharp Climate Shift." *Los Angeles Times*, July 13, 2001, A-1.

———. "Scientists Now Fear 'Abrupt' Global Warming Changes: Severe and 'Unwelcome' Shifts Could Come in Decades, Not Centuries, a National Academy Says in an Alert." *Los Angeles Times*, December 12, 2001, A-30.

McNeill, J. R. *Something New under the Sun: An Environmental History of the Twentieth-Century World.* New York: W.W. Norton, 2000.

Melillo, J. M., P. A. Steudler, J. D. Aber, K. Newkirk, H. Lux, F. P. Bowles, et al. "Soil Warming and Carbon-Cycle Feedbacks to the Climate System." *Science* 298 (December 13, 2002): 2173–2176.

Menon, Surabi, James Hansen, Larissa Nazarenko, and Yunfeng Luo. "Climate Effects of Black Carbon Aerosols in China and India." *Science* 297 (September 27, 2002): 2250–2253.

Michaels, Patrick J. "'Day After Tomorrow': A Lot of Hot Air." *USA Today*, May 25, 2004, 21-A.

Mieszkowski, Katharine. "Bush: Global Warming Is Just Hot Air." Salon.com, September 10, 2004. (Lexis).

Minnis, Patrick, J. Kirk Ayres, Rabindra Palikonda, and Dung Phan. "Contrails, Cirrus Trends, and Climate." *Journal of Climate*, April 5, 2004, 1671–1685.

"Most Serious Greenhouse Gas Is Increasing, International Study Finds." *Science Daily*, April 27, 2001. www.sciencedaily.com/releases/2001/04/010427071 254.htm.

Munro, Margaret. "Earth's 'Big Burp' Triggered Warming: Prehistoric Release of Methane a Cautionary Tale for Today." *Edmonton Journal* (Canada), June 3, 2004, A-10.

Nance, John J. *What Goes Up: The Global Assault on Our Atmosphere.* New York: William Morrow and Co., 1991.

National Academy of Sciences. *Policy Implications of Greenhouse Warming.* Washington, D.C.: National Academy Press, 1991.

Nemani, Ramakrishna R., Charles D. Keeling, Hirofumi Hashimoto, William M. Jolly, Stephen C. Piper, Compton J. Tucker, et al. "Climate-Driven

Increases in Global Terrestrial Net Primary Production from 1982 to 1999." *Science* 300 (June 6, 2003): 1560–1563.

Nesmith, Jeff. "Dirty Snow Spurs Global Warming: Study Says Soot Blocks Reflection, Hurries Melting." *Atlanta Journal-Constitution*, December 23, 2003, 3-A.

———. "Is the Earth too Hot? A New Study Says No, but Then It Was Funded by Big Oil Companies." *Atlanta Journal and Constitution* in *Hamilton Spectator* (Canada), May 30, 2003. (Lexis).

"New Climate Model Predicts Greater 21st Century Warming." AScribe Newswire, May 19, 2003. (Lexis).

"New Look at Satellite Data Supports Global Warming Trend." AScribe Newswire, April 30, 2003. (Lexis).

Newman, Paul A. "Preserving Earth's Stratosphere." *Mechanical Engineering*, October 1998. www.memagazine.org/backissues/october98/features/stratos/stratos.html.

"New NASA Satellite Sensor and Field Experiment Shows Aerosols Cool the Surface but Warm the Atmosphere." National Aeronautics and Space Administration public information release, August 15, 2001. http://earthobservatory.nasa.gov/Newsroom/MediaResources/Indian_Ocean_Experiment/indoex_release.html.

"North Pole Had Sub-Tropical Seas Because of Global Warming." Agence France Presse, September 7, 2004. (Lexis).

O'Carroll, Cynthia. "NASA Blames Greenhouse Gases for Wintertime Warming." *UniSci*, April 24, 2001. http://unisci.com/stories/20012/0424011.htm.

Oppenheimer, Michael, and Robert H. Boyle. *Dead Heat: The Race against the Greenhouse Effect*. New York: Basic Books, 1990.

Pacala, S. W., G. C. Hurtt, D. Baker, P. Peylin, R. A. Houghton, R. A. Birdsey, et al. "Consistent Land- and Atmosphere-Based U.S. Carbon Sink Estimates." *Science* 292 (June 22, 2001): 2316–2320.

Page, Susan E., Florian Siegert, John O. Rieley, Hans-Dieter V. Boehm, Adi Jaya, and Suwido Limin. "The Amount of Carbon Released from Peat and Forest Fires in Indonesia during 1997." *Nature* 420 (November 7, 2002): 61–65.

Parsons, Michael L. *Global Warming: The Truth Behind the Myth*. New York: Plenum Press/Insight, 1995.

Pearce, Fred. "Massive Peat Burn Is Speeding Climate Change." New Scientist.com, November 3, 2004. www.newscientist.com/news/news.jsp?id=ns99996613.

Pearson, Paul N., Peter W. Ditchfield, Joyce Singano, Katherine G. Harcourt-Brown, Christopher J. Nicholas, Richard K. Olsson, et al. "Warm Tropical

Sea-Surface Temperatures in the Late Cretaceous and Eocene Epochs." *Nature* 413 (October 4, 2001): 481–487.

Pianin, Eric. "Greenhouse Gases Decrease: Experts Cite U.S. Economic Decline, Warm Winter." *Washington Post*, December 21, 2002, A-2. www .newscientist.com/news/news.jsp?id=ns99996613.

Pielke, Roger. "Land Use Changes and Climate Change." *Philosophical Transactions: Mathematical, Physical & Engineering Sciences* (Journal of The Royal Society of London), August, 2002.

"Planting Northern Forests Would Increase Global Warming."*New Scientist*, Press release, July 11, 2001. www.newscientist.com/news/news.jsp?id=ns 99991003.

Polakovic, Gary. "Airborne Soot Is Significant Factor in Global Warming, Study Says." *Los Angeles Times*, May 15, 2003, A-30.

"Pollution Adds to Global Warming." Via Excite! Data feed, 3:23 a.m. ET, Associated Press, October 26, 2000. http://apple.excite.com.

Pomerance, Rafe. "The Dangers from Climate Warming: A Public Awakening." In *The Challenge of Global Warming*, ed. Edwin Abrahamson, 259–269. Washington, D.C.: Island Press, 1989.

Powlson, David. "Will Soil Amplify Climate Change?" *Nature* 433 (January 20, 2005): 204–5.

Radford, Tim. "Scientists Discover the Harbinger of Drought: Subtle Temperature Changes in Tropical Seas May Trigger Northern Hemisphere's Long, Dry Spells." *Guardian* (London), January 31, 2003, 18.

———. "World May Be Warming Up Even Faster: Climate Scientists Warn New Forests Would Make Effects Worse." *Guardian* (London), November 9, 2000, 10.

Radowitz, John von. "Calmer Sun Could Counteract Global Warming." Press Association, October 5, 2003. (Lexis).

Ramanathan, V. "Observed Increases in Greenhouse Gases and Predicted Climatic Changes." In *The Challenge of Global Warming*, ed. Edwin Abrahamson, 239–247. Washington, D.C.: Island Press, 1989.

Ramanathan, V., P. J. Crutzen, J. T. Kiehl, and D. Rosenfeld. "Aerosols, Climate, and the Hydrological Cycle." *Science* 294 (December 7, 2001): 2119–2124.

Randall, C. E., V. L. Harvey, G. L. Manney, Y. Orsolini, M. Codrescu, C. Sioris, et al. "Stratospheric Effects of Energetic Particle Precipitation in 2003–2004." *Geophysical Research Letters* 32 (March 2, 2005). L05802, doi: 10.1029/2004GL022003.

Raymond, Peter A., and Jonathan J. Cole. "Increase in the Export of Alkalinity from North America's Largest River." *Science* 301(July 4, 2003): 88–91.

Regalado, Antonio. "Skeptics on Arming Are Criticized." *Wall Street Journal*, July 31, 2003, A-3, A-4.

"Research Casts Doubt on China's Pollution Claims." *Washington Post*, August 15, 2001, A-16. www.washingtonpost.com/wp-dyn/articles/A10645-2001 Aug14.html.

Revkin, Andrew C. "Both Sides Now: New Way That Clouds May Cool." *New York Times*, June 19, 2001, F-4.

———. "Bush Versus the Laureates: How Science Becomes a Partisan." *New York Times*, October 19, 2004, F-1.

Rex, Markus, P. von der Gathen, Alfred Wegener, R. J. Salawitch, N. R. P. Harris, M. P. Chipperfield, et al. "Arctic Ozone Loss and Climate Change." *Geophysical Research Letters* 31 (March 10, 2004). www.eurekalert.org/pub_releases/2004-03/agu-ajho31004.php.

Richardson, Michael. "Indonesian Peat Fires Stoke Rise of Pollution." *International Herald-Tribune*, December 13, 2002, 5.

Robock, Alan. "Pinatubo Eruption: The Climatic Aftermath." *Science* 295 (February 15, 2002): 1242–1244.

"Rock Measurements Suggest Warming Is Global." Environment News Service, April 16, 2002. http://ens-news.com/ens/apr2002/2002L-04-16-09.html.

Rowland, Sherwood, and Mario Molina. "Stratospheric Sink for Chlorofluoromethanes: Chlorine Atom-Catalyzed Destruction of Ozone." *Nature* 249 (June 28, 1974): 810–812.

Russell-Jones, Robin. "Ozone in Peril." Letter to the editor, *Independent* (London), December 7, 2000, 2.

Sabine, Christopher L., Richard A. Feely, Nicolas Gruber, Robert M. Key, Ki-tack Lee, John L. Bullister, et al. "The Oceanic Sink for Anthropogenic CO_2." *Science* 305 (July 16, 2004): 367–371.

Sato, Makiko, James Hansen, Dorothy Koch, Andrew Lacis, Reto Ruedy, Oleg Dubovik, et al. "Global Atmospheric Black Carbon Inferred from AERONET." *Proceedings of the National Academy of Sciences* 100 (11) (May 27, 2003): 6319–6324.

Schiermeier, Quirin. "Gas Leak: Global Warming Isn't a New Phenomenon—Sea-Bed Emissions of Methane Caused Temperatures to Soar in Our Geological Past, but No One Is Sure What Triggered the Release." *Nature* 423 (June 12, 2003): 681–682.

Schimel, D. S., J. I. House, K. A. Hibbard, P. Bousquet, P. Cials, P. Peylin, et al. "Recent Patterns and Mechanisms of Carbon Exchange by Terrestrial Ecosystems." *Nature* 414 (November 8, 2001): 169–172.

Schimel, David, and David Baker. "Carbon Cycle: The Wildlife Factor." *Nature* 420 (November 7, 2002): 29–30.

Schleck, Dave. "High Fliers May Be Creating Clouds: Global Warming May Be Worsened by Contrails from Aircraft." *Montreal Gazette* (Canada), July 18, 2004, D-6.

Schlesinger, W. H., and J. Lichter. "Limited Carbon Storage in Soil and Litter of Experimental Forest Plots Under Increased Atmospheric CO_2." *Nature* 411 (May 24, 2001): 466–469.

Schneider, Stephen H. *Global Warming: Are We Entering the Greenhouse Century?* San Francisco: Sierra Club Books, 1989.

Schoffner, Chuck. "NASA: Bush Stifles Global Warming Evidence." Associated Press, October 27, 2004. (Lexis).

Schrope, Mark. "Successes in Fight to Save Ozone Layer Could Close Holes by 2050." *Nature* 408 (December 7, 2000): 627.

Semple, Robert B. "A Film That Could Warm Up the Debate on Global Warming." Editorial Observer. *New York Times*, May 27, 2004, http://nytimes.com/2004/05/27/opinion/27THU3.html.

"Severe Loss to Arctic Ozone." British Broadcasting Corporation News, April 5, 2000. http://news.bbc.co.uk/hi/english/sci/tech/newsid_702000/702388.stm.

Shindell, D. T., G. A. Schmidt, R. L. Miller, and D. Rind. "Northern Hemisphere Winter Climate Response to Greenhouse Gas, Ozone, Solar, and Volcanic Forcing." *Journal of Geophysical Research* 106 (2001): 7193–7210.

Shwartz, Mark. "New Study Reveals a Major Cause of Global Warming—Ordinary Soot." *Stanford University Departmental News*, February 7, 2001. www.stanford.edu/dept/news/.

Silver, Cheryl Simon, and Ruth S. DeFries. *One Earth, One Future: Our Changing Global Environment.* Washington, D.C.: National Academy Press, 1990.

Siegenthaler, Urs, Thomas F. Stocker, Eric Monnin, Jakob Schwander, Bernhard Stauffer, Dominique Raynaud, Jean-Marc Barnola, Hubertus Fischer, Valèrie Masson-Delmotte, and Jean Jouzel. "Stable Carbon Cycle Climate Relationship during the Late Pleistocene." *Science* 310 (November 25, 2005): 1313–1317.

Siskind, David E., Stephen D. Eckermann, and Michael E. Summers, eds. *Atmospheric Science across the Stratopause*, Geophysical Monograph 123. Washington, D.C.: American Geophysical Union, 2000.

Solanki, S. K., I. G. Usoskin, B. Kromer, M. Schussler, and J. Beer. "Unusual Activity of the Sun during Recent Decades Compared to the Previous 11,000 Years." *Nature* 431 (October 28, 2004): 1084–1086.

Sorensen, Eric. "The Letter We Can't See: Atmospheric Carbon Is One of the Worst Culprits in Global Warming." *Seattle Times*, April 22, 2001, A-1.

Spahni, Renato, Jèrùme Chappellaz, Thomas F. Stocker, Laetitia Loulergue, Gregor Hausammann, Kenji Kawamura, Jacqueline Flückiger, Jakob

Schwander, Dominique Raynaud, Valèrie Masson-Delmotte, and Jean Jouzel. "Atmospheric Methane and Nitrous Oxide of the Late Pleistocene from Antarctic Ice Cores." *Science* 310 (November 25, 2005): 1317–1321.

Speth, James Gustave. *Red Sky at Morning: America and the Crisis of the Global Environment*. New Haven, CT: Yale University Press, 2004.

Spotts, Peter N. "Trees No Savior for Global Warming." *Christian Science Monitor*, May 25, 2001.

"Spray Cans Warming the Planet, One Dust-Busting Puff at a Time." *International Herald-Tribune* in *Herald Asahi* (Tokyo), May 27, 2004. (Lexis).

Stainforth, D. A., T. Aina, G. Christensen, M. Collins, N. Faull, D. J. Frame, et al. "Uncertainty in Predictions of the Climate Response to Rising Levels of Greenhouse Gases." *Nature* 433 (January 27, 2005): 403–406.

Stevens, William K. *The Change in the Weather: People, Weather, and the Science of Climate*. New York: Delacorte Press, 1999.

"Storm Warning." *Boston Globe*, May 23, 2004, N-17.

Streets, David G., Kejun Jiang, Xiulian Hu, Jonathan E. Sinton, Xiao-Quan Zhang, Deying Xu, et al. "Recent Reductions in China's Greenhouse Gas Emissions." *Science* 294 (November 30, 2001): 1835–1837.

"Study of Ancient Air Bubbles Raises Concern about Today's Greenhouse Gases." Omaha *World-Herald*, November 25, 2005, 4-A.

"Study Shows Soil Warming May Stimulate Carbon Storage in Some Forests; Effect Would Slow Rate of Climate Change." AScribe Newsletter, December 12, 2002. (Lexis).

Svensen, Henrik, Sverre Planke, Anders Malthe-Sorenssen, Bjorn Jamtveit, Reidun Myklebust, Torfinn Rasmussen Eidem, et al. "Release of Methane from a Volcanic Basin as a Mechanism for Initial Eocene Global Warming." *Nature* 429 (June 3, 2004): 542–545.

"Swedish Bogs Flooding Atmosphere with Methane: Thawing Sub-Arctic Permafrost Increases Greenhouse Gas Emissions." American Geophysical Union, February 10, 2004. www.scienceblog.com/community/article2366.html.

Tabazadeh, A., E. J. Jensen, O. B. Toon, K. Drdla, and M. R. Schoeberl. "Role of the Stratospheric Polar Freezing Belt in Denitrification." *Science* 291(March 30, 2001): 2591–2594.

Takahashi, Taro. "The Fate of Industrial Carbon Dioxide." *Science* 305 (July 16, 2004): 352–353.

Taylor, James M. "Hollywood's Fake Take on Global Warming." *Boston Globe*, June 1, 2004, A-11.

Thompson, David W. J., and John M. Wallace. "Regional Climate Impacts of the Northern Hemisphere Annular Mode." *Science* 293 (July 6, 2001): 85–89.

Tolbert, Margaret A., and Owen B. Toon. "Solving the P[olar] S[tratospheric] C[loud] Mystery." *Science* 292 (April 6, 2001): 61–63.

Toner, Mike. "Drought May Signal World Warming Trend." *Atlanta Journal and Constitution*, January 31, 2003, 4-A.

"Top U.S. Newspapers' Focus on Balance Led to Skewed Coverage of Global Warming, Analysis Reveals." AScribe Newsletter, August 25, 2004. (Lexis).

Travis, Davis J., Andrew M. Carleton, and Ryan G. Lauritsen. "Climatology: Contrails Reduce Daily Temperature Range." *Nature* 418 (August 8, 2002): 593–594.

Tudhope, Alexander W., Colin P. Chilcott, Malcolm T. McCulloch, Edward R. Cook, John Chappell, Robert M. Ellam, et al. "Variability in the El Niño-Southern Oscillation through a Glacial-Interglacial Cycle." *Science* 291 (February 23, 2001): 1511–1517.

Unger, Mike. "Global Warming Hits Big Screen." *Annapolis Capital*, May 28, 2004, A-1.

Veizer, Jan, Yves Godderis, and Louis M. Francois. "Evidence for Decoupling of Atmospheric CO_2 and Global Climate during the Phanerozoic Eon." *Nature* 408 (December 7, 2000): 698–701.

Victor, Davis G. *Climate Change: Debating America's Policy Options.* New York: Council on Foreign Relations, 2004.

Vidal, John. "You Thought It Was Wet? Wait until the Asian Brown Cloud Hits Town: Extreme Weather Set to Worsen through Pollution and El Niño; Cloud with No Silver Lining." *Guardian* (London), August 12, 2002, 3.

Vinnikov, Konstantin Y., and Norman C. Grody. "Global Warming Trend of Mean Tropospheric Temperature Observed by Satellites." *Science* 302 (October 10, 2003): 269–272.

"Warming Tropics Show Reduced Cloud Cover." Environment News Service, February 1, 2002. http://ens-news.com/ens/feb2002/2002L-02-01-09.html.

Whalley, John, and Randall Wigle. "The International Incidence of Carbon Taxes." Paper presented at a conference on economic policy responses to global warming in Rome, Italy, September 1990.

White, James C. "Do I Hear a Million?" *Science* 304 (June 11, 2004): 1609–1610.

Wielicki, Bruce A., Takmeng Wong, Richard P. Allan, Anthony Slingo, Jeffrey T. Kiehl, Brian J. Soden, et al. "Evidence for Large Decadal Variability in the Tropical Mean Radiative Energy Budget." *Science* 295 (February 1, 2002): 841–844.

"Wildfires Add Carbon to the Atmosphere." Environment News Service, December 9, 2002. http://ens-news.com/ens/dec2002/2002-12-09-09.asp.

Wofsy, Steven C. "Where Has All the Carbon Gone?" *Science* 292 (June 22, 2001): 2261–2263.

Wolff, Eric W., Laurent Augustin, Carlo Barbante, Piers R. F. Barnes, Jean Marc Barnola, Matthias Bigler, et al. "Eight Glacial Cycles from an Antarctic Ice Core." *Nature* 429 (June 10, 2004): 623–627.

Woodwell, George M., and Fred T. MacKenzie, eds. *Biotic Feedbacks in the Global Climate System: Will the Warming Feed the Warming?* New York: Oxford University Press, 1995.

"Yale University: Vast Majority of Americans Believe Global Warming Is 'Serious Problem.'" M2 Presswire, May 29, 2004. (Lexis).

Yam, Philip. "A Less Carbonated Earth." *Scientific American*, October 1999, 32.

Younge, Gary. "Bush U-turn on Climate Change Wins Few Friends." *Guardian* (London), August 27, 2004, 18.

II THE WEATHER NOW—AND IN 2100

INTRODUCTION

Since worldwide weather record keeping began about 1880 until 2004, the eighteen warmest years all occurred after 1980. Nine of the ten warmest years on record had taken place after 1990. The ten warmest years on record all had occurred since 1987. In England, four of the five warmest years in the 330-year-old Central England Temperature Record occurred after 1990. By 2003, the world had seen twenty-six consecutive years of above-average temperatures, based on thirty-year moving averages. The last time Earth was cooler than long-term averages had been 1976.

While warming may be global in character, it is also sometimes highly variable from region to region. Some areas (an example is the United States' southeastern states) experienced little or no temperature increase during the late twentieth century. Trends in mean annual air temperature for 1950 to 1998 indicate three areas of particularly rapid regional warming, all at high latitudes: northwestern North America (including Alaska), an area around the Siberian Plateau, and the Antarctic Peninsula (Vaughan et al. 2001, 1777). Rapid regional warming around the Antarctic Peninsula has led to loss of seven ice shelves during the last fifty years.

As temperatures have increased, a United Nations report for the August 2002 Earth Summit said that fossil fuel consumption and carbon dioxide emissions continued to rise during the 1990s, particularly in Asia and North America. Other signals of climate change linked to global warming also were more apparent, including more-frequent and intense droughts in parts of Asia and Africa. Sea levels also were rising (Lederer 2002). Case-by-case examples of a rapidly warming Earth were plentiful in the early years of the new millennium. Warming often was accompanied by increasing extremes in precipitation; some areas were increasingly subject to alternating droughts and deluges as warming distorted the hydrological cycle.

In *Abrupt Climate Change: Inevitable Surprises* (2002), Richard B. Alley, an expert on abrupt climate change at Pennsylvania State University, wrote that climate could change rapidly (and has in the past). "When gradual causes push the Earth system across a threshold. Just as the slowly increasing pressure of a finger eventually flips a switch and turns on a light ..." (Alley 2002, v). The temperature record for Greenland, according to Alley's research, more resembles a jagged row of very sharp teeth than a gradual passage from one climatic epoch to another. According to Alley, "Model projections of global warming find increased global precipitation, increased variability in precipitation, and summertime drying in many continental interiors, including grain belt regions. Such changes might produce more floods and more droughts" (p. 114).

6 WATCHING THE THERMOMETER

CLIMATE CHANGE IN THE PRESENT TENSE

During February 2001, the Intergovernmental Panel on Climate Change (IPCC) switched its focus to the present from the future. For the first time, it stated with "high confidence" that recent changes in the world's climate have had "discernible" impacts on physical and biological systems. More than 100 governments represented on the IPCC concluded a meeting in Geneva, Switzerland, by accepting a report stating that a warming global environment already was having worldwide impact (Lean 2001, 12).

The most acute consequences of rapid climate change are expected to afflict impoverished countries in the third world that have the least capacity to adapt. Much of Africa, for example, is highly vulnerable to climate changes that can affect water resources, food production, and the expansion of deserts and cause more-frequent outbreaks of diseases of cholera. The report lists a number of small island-states in the Pacific and Indian oceans and the Caribbean Sea that are threatened by climate change, where unique cultural and conservation sites already have been destroyed. Glaciers in tropical regions such as the Himalayas are particularly threatened by climate change, according to the IPCC. Himalayan glaciers are the major source of water for the Ganges and Indus rivers, on which 500 million people, just under one-tenth of the world's population, depend (see Chapter 12, "Mountain Glaciers in Retreat").

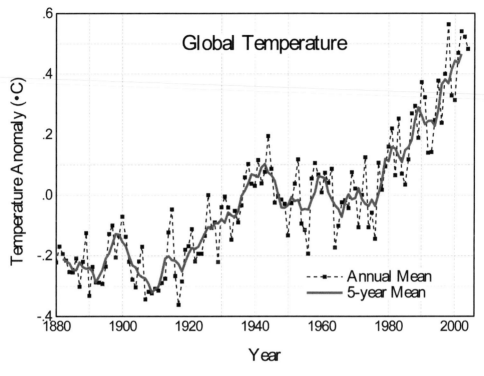

Trend of global annual surface temperature, 1880–2000, relative to the 1951–1980 mean. Courtesy of the Goddard Institute of Space Studies/NASA.

PERSISTENT WARMTH WORLDWIDE

Average temperatures worldwide rose rapidly during the late 1990s, peaking during 1998 as the result of a strong El Niño weather event. Temperatures then subsided slightly but began to climb once again by 2001, even in the absence of a strong El Niño pattern. "The fact that 2002 is almost as warm as the unusual warmth of 1998 is confirmation that the underlying global warming trend is continuing," said James Hansen of NASA's Goddard Institute for Space Studies in New York City (McFarling 2002, A-5).

Although the year 2000 ended on a nippy note in much of North America, National Oceanic and Atmospheric Administration (NOAA) figures showed that temperatures for the year as a whole were much higher than usual in the United States, with an average of 54.1 degrees F, well above the long-term average of 52.8 degrees F, making 2000 the seventh-warmest year in 106 years of record keeping. The year 2000's annual figure was down from the 54.5 degrees F average of 1999 and from 1998's all-time record of 54.9 degrees F, according to NOAA's National Climatic Data Center in Asheville, North Carolina.

After a slight retreat in 1999 and 2000, temperatures were again hugging record highs in 2001, 2002, and 2003, aided by a weak El Niño event during most of 2002. Clearly, the underlying average (absent El Niño) seemed to be rising persistently. Hansen and colleagues observed that during 2001, global temperatures averaged the second highest in recorded history (which included more than a century of instrumental data), exceeded only by 1998. "The global warmth of 2001 is particularly meaningful because it occurred at a phase of the Southern Oscillation in which the tropical Pacific Ocean was cool," Hansen and colleagues wrote in a letter to *Science* (Hansen 2002, 275). The warming trend persisted. Worldwide temperatures in January, February, and March of 2002 were the warmest ever recorded for those months (Toner 2002, 12-A). The first seven months of 2002 averaged 58.37 degrees F globally, according to the Goddard Institute, a fraction of a degree less than the record of 58.44 degrees F set during the same period in 1998.

The year 2003 was the third warmest since modern records have been kept worldwide, after 1998 and 2002. The year 2003 also marked the twenty-seventh consecutive annual cycle that temperatures had exceeded historical averages. Worldwide temperatures in September 2003 were the highest on record, according to the U.S. National Climatic Data Center. The year 2004 continued the trend, finishing as the fourth-warmest annual cycle on the instrumental record.

The year 2005 was the second-warmest worldwide on the instrumental record (since 1861), and the warmest in the Northern Hemisphere, according to the World Meteorological Organization and the U.S. National Climatic Data Center. The average temperature worldwide was 58.1 degrees F, one-tenth of a degree shy of the 1998 record of 58.2. The 1998 record at the time was much warmer than any previous year, and was aided by a very strong El Niño event, which usually raises temperatures. The 2005 average, by contrast, occurred with no aid from an El Niño event. Warming was strongest over the Arctic, where loss of sea ice is changing reflectivity to aid future warming (O'Driscoll 2005, 2-A).

Winters became shorter in the United States during the last half of the twentieth century. The date of the last spring freeze in the United States occurred earlier during the period from 1948 to 1999 than in previous years, according to David Easterling, a climatologist with NOAA's National Climatic Data Center. The annual number of frost-free days grew over that fifty-two-year period, Easterling noted. He presented his research on January 16, 2001, at the annual meeting of the American Meteorological Society in Albuquerque.

The date of the first frost in autumn showed little change, according to Easterling, but the average date of the last spring freeze showed significant movement to an earlier date. The IPCC's Third Assessment indicated that rising temperatures have extended the "freeze-free" season in many mid- and high-latitude regions. Snow cover decreased 10 percent between the 1960s and the 1990s (Smith 2001, 7).

The abrupt nature of warming during the last few decades has been illustrated by an analysis of proxies for temperature records during the last 2,000 years, using information inferred from tree rings, stalagmites, seabed layers, and other evidence in the Northern Hemisphere. This study indicated that previous peaks of warming, particularly during medieval times about 1,000 years ago, were as warm as the twentieth-century-average, "but that no spikes in the last 2,000 years matched the warming since 1990" (Revkin 2005).

Anders Moberg and colleagues wrote, "According to our recon-struction, high temperatures—similar to those observed in the twentieth century before 1990—occurred around A.D. 1000 to 1100, and minimum temperatures that are about 0.7 degrees C. below the average of 1961–90 occurred around A.D. 1600. Large natural variability in the past sug-gests an important role of natural multicentennial variability that is likely to continue" (Moberg et al. 2005, 613). This reconstruction indicates that temperatures during the last 1,000 years ranged both higher and lower than previously believed, accounting, for example, for unusual warmth in Greenland at the time of Viking expansion, as well as the intense cold of the "little ice age" about 500 years later. Moberg, of Sweden's Stockholm University, said that natural influences on climate could either amplify or mask human-caused warming in years to come. However, said Moberg, this study "should not be a fuel for greenhouse skeptics in their arguments" (Revkin 2005). He added that ample evidence exists that recent warming was now exceeds natural limits.

EXTREMES OF HEAT AND COLD AROUND THE WORLD

While weather has been generally warmer worldwide, reports of heat waves have been punctuated by occasional record cold and deep snowfall during unusual times of the year. A weeks-long heat wave in southern India killed at least 1,000 people during mid-May 2002 in Andhra Pradesh state. Temperatures in the shade rose as high as 124 degrees F, killing birds in trees and making many homes uninhabitable. By May 22, the death toll in Andhra Pradesh was the highest in any Indian heat wave—until the next year. Most of those who died were elderly and poor. People who worked outside, such as farm laborers and rickshaw drivers, died in large numbers.

On May 23, one day later, Cut Bank, Montana, experienced enough snow and wind to pile drifts four feet high, and a day later, wet snow fell in Denver, Colorado. In Canada, the same spring was the fifth coldest in half a century, breaking a string of five years during which every season had been warmer than average (Spears 2002, A-8). At the same time, the growing season in Albuquerque, New Mexico, was reported to have lengthened by two weeks between 1931 and 2002, to an average of 204 days, according to the Albuquerque office of the National Weather Service, which attributed the change to the urban heat-island effect and global warming (Fleck 2002, A-1).

During January 2003, extreme cold moved into some of the same areas of India that had experienced searing heat the previous May. The cold wave killed at least 1,800 people, mainly from exposure, in Bangladesh and India, where many people sleep outside on the earth with no protective clothing, as well as in Nepal. At about the same time, Vietnam lost a third of its rice crop to unusual cold. The period of May through August of 2004 was the coldest on record in Winnipeg, Manitoba, since record keeping began in 1872. On the roster of cold surprises in a warming world, residents of Galveston and Corpus Christi, Texas, will long remember their white Christmas of 2004. Between those two cities (and a few miles inland), Victoria, Texas, was buried under 10 inches of wet snow, its first white Christmas in eighty-six years. Brownsville, Texas, received 1.5 inches of snow the same day, the first measurable snow there since 1895. A Brownsville snowball sold for $92 on eBay ("How About" 2005, C-16).

Mikhail Koslov and Natalia G. Berlina, analyzing records from a Lapland reserve on the Kola Peninsula of Russia during 1930 and 1998 found "a decline in the length of the snow-free and ice-free periods by 15 to 20 days due to both delayed spring and advanced autumn/winter" (Koslov and Berlina 2002, 387). Emissions of sulfur dioxide from a nearby industrial plant might have contributed to the cooling, which was associated with a snowfall increase of more than 40 percent during the same period. After a hot summer and a November that was 4.5 degrees F above average, Omaha in 2005 was afflicted with their coldest first week of December on record. Omaha's temperature averaged 9.3 degrees F December 1–8, compared to the previous record of 13.1 in 1913. By Christmas, however, temperatures in Omaha had risen to 15 to 20 degrees *above* average several days in a row.

Berlin experienced its coldest October on record in 2002, with earlier-than-usual snows. Omaha, Nebraska, also experienced a notably cool October in 2002, with early snow and a monthly average temperature 6.6 degrees F below average. On May 6, 2003, residents of St. Albert, Alberta (near Edmonton) looked out their windows at a foot of snow. "How long is this global warming going to take," complained St. Albert resident Tim Enger in the *Edmonton Journal*. "I'm waiting" (Enger 2003, A-19).

The following summer, in 2003, more than 1,400 people died of heat stress as temperatures rose as high as 122 degrees F in India during three weeks in late May and early June, before the annual monsoon began. The same area also suffered its worst drought in at least twenty years. An analysis by Uday Shankar De, a geophysicist and former research chief at the Indian Meteorological Department in Pune, suggested that the number of hot, dry days in India during May and June had increased steadily over the previous two decades. De and colleagues attributed the trend to global climate change. "The increase in extreme events, such as the heat wave, is linked to global warming," asserted J. Srinivasan, a professor of atmospheric sciences at the Indian Institute of Science in Bangalore (Jayaraman 2003, 673). India's heat wave of 2003 intensified as hot, dry air flowed eastward from deserts in Iran; by May 2003, according to satellite data supplied by the U.S. National Oceanic and Atmospheric Administration, two-thirds of vegetation in India was under "severe" or "extreme" drought conditions (p. 673).

HUMAN INFLUENCES NOW DOMINATE CLIMATE CHANGE

According to atmospheric scientists Thomas Karl and Kevin Trenberth, writing in *Science*, natural variations ceased being the main determinants of Earth's climate about 1950. Since then, "Human influences have been the dominant detectable influence on climate change" (Karl and Trenberth 2003, 1720). The human overload of greenhouse gases in the atmosphere has become the main (although not the only) driver of climate. Among human causes, land-use changes and urbanization also are adding heat to the atmosphere, according to Karl and Trenberth. The two scientists' models indicate a 90 percent probability that temperatures worldwide could rise between 3.1 and 8.9 degrees F by the year 2100. They concluded:

> There is still considerable uncertainty about the rates of change that can be expected, but it is clear that these changes will be

increasingly manifested in important and tangible ways, such as changes in extremes of temperature and precipitation, decreases in seasonal and perennial snow and ice extent, and sea-level rise. Anthropogenic climate change is now likely to continue for many centuries. We are venturing into the unknown with climate, and its associated impacts could be quite disruptive. (Karl and Trenberth 2003, 1719)

In the United States, the period of November 1, 2001, through January 31, 2002, was the warmest on record (4.3 degrees F above average); globally, January 2002 was the warmest January on record. The average U.S. temperature over those three months was 39.94 degrees—4.3 degrees more than the three-month average for the previous 106 years, according to the National Climatic Data Center in Asheville, North Carolina. The old record for that period, 39.63 degrees F, had been recorded only two years earlier. All forty-eight of the contiguous states were warmer than average in November, December, and January. For twenty-three of those states, the period was the warmest or second warmest on record, according to the National Center for Atmospheric Research in Boulder, Colorado. Five states—Minnesota, Wisconsin, Iowa, Massachusetts, and Vermont—set records for warmth. The coolest state was California, and it was still warmer than in two-thirds of its past early winters.

"Studying these annual temperature data, one gets the unmistakable feeling that temperature is rising and that the rise is gaining momentum," said Lester R. Brown, an economist and president of the Earth Policy Institute in Washington, D.C. (McFarling [December 12] 2002, A-5). The string of warmer years provides strong evidence that humans are in large part to blame for changing the climate, said Peter Frumhoff, an ecologist and senior scientist with the Union of Concerned Scientists in Cambridge, Massachusetts. "It's important we pay attention to this drumbeat of evidence as the signal of human impact starts to emerge from the noise of natural climate patterns," he said (p. A-5).

A SUMMERY MORNING IN EARLY DECEMBER

At street level, on a day-to-day basis, surges of unusual warmth could be very dramatic. In Anchorage, Alaska, during November 2002, for twenty days the low temperature was higher than the average high. The city ended the month with no snow on the ground, a most unusual event. During October and November, Anchorage had less than 10 percent of its average snowfall. At the same time, the U.S. Northeast and Mid-Atlantic regions experienced repeated snowfalls. More snow was recorded that winter in Richmond, Virginia, than in Anchorage. A snow drought also afflicted Fairbanks, Alaska. For the first time in recorded history, Fairbanks, during the winter of 2002, did not record a single nighttime temperature below −40 degrees F.

The period through November and into early December 2001 was very warm, especially in the northern and eastern sections of the United States. John Kahionhes Fadden, a Mohawk culture bearer, reported on December 2 that there was no snow at his house high in the Adirondack Mountains of far upstate New York, which is very unusual for that date. Buffalo, New York, reported no snow in November for the first time since records had been kept, and Marquette, Michigan, reported a record average high temperature, 39.9 degrees F, which was 10.4 degrees F above the usual average. (Weather is still variable, of course. Two years later, Fadden reported copious snow and temperature as low as minus 36 degrees F.)

The average temperature in Omaha, Nebraska, for November 2001 was 49.5 degrees F—10.4 F degrees above the usual average and an all-time record high (dating from 1871), exceeding the previous record set two years earlier by a remarkable 2.4 degrees F. Some trees were putting out buds in late fall. The local newspaper contained no speculation about global warming in its page-one article on the unusual warmth. The headline on the piece bid readers to "enjoy" the warm spell (Gaardner 2001, A-1). On December 5, 2001, in Omaha, the temperature was 65 degrees at 6 a.m., a record daily high before the sun came up, feeling like August during the first week of December.

The warm, humid early morning was followed by thunder, lightning, and a torrential downpour, portending passage of a summer-style cold front. The December 6, 2001, edition of the *Omaha World-Herald* published a page-one story headlined, "Warm Weather a Mixed Blessing." Although the headline suggested news of climate change, the news report, however, concentrated on flagging sales of snowmobiles in a season that had been, to that date, utterly bereft of snow.

The unusual weather was not restricted to Omaha, of course. Josh Rubin commented in the *Toronto Star*: "A rose by any other name might smell as sweet. But blooming in Toronto, in December? Roses are blooming, bulbs are sprouting, golfers are still hitting the links and climatologists are puzzled" (Rubin 2001, B-2). "People are coming in and asking what to do about their roses still blooming, and wondering if it's safe. Some of their trees are starting to bud. It's pretty weird for this time of year," said John Manning, manager of White Rose Nursery in East York (p. B-2). At Stouffville's Rolling Hills Golf Club, manager John Finlayson said the phone had been ringing off the hook with golfers requesting tee times. "It's pretty much been the busiest it's been all year," said Finlayson, who kept the course open well past the usual end of the season. "We'd normally close down some time in November, and it wouldn't usually be so busy late in the season," said Finlayson (p. B-2).

At about the same time, *Boston Globe* columnist Derrick Z. Jackson wrote:

> On Thanksgiving morning here in the cradle of the new North, I went to my backyard to clip a full bunch of cilantro for my guacamole. Earlier that week, I snipped four nice-sized eggplants from the vine, as well as my last jalapeno peppers to smoke into the chipoltes that would go into the guacamole. More fresh cilantro is still poking up out of the ground. It is not soaring to reach the summer's sun, but its greens are still vibrant and the taste still a pungent surprise. A garden in New England in December, three weeks before the winter solstice, used to be a non sequitur....

Being this excited about seeing my garden grow, it is hard to see the madness behind the miracle. Behind the heat of my jalapenos, I know there is supposed to be a throb in my head: "glo... bal... war... ming...glo-bal... war-ming... globalwarming." (Jackson 2001, A-23)

Jackson wrote that the New England Regional Assessment Group, an arm of the federal U.S. Global Change Research Program, had forecast climatic warming by 2090 that could give Boston the late-twentieth-century climate of Richmond, Virginia, or Atlanta, Georgia. "That rise could mean a lot more than cilantro and eggplant. It might mean more smog, acid rain, and red tides. The sugar maples of our spring syrup and our fall foliage could be in trouble. And those poor beach-front millionaires.... If Boston becomes Atlanta, what does Atlanta become? The Sahara? The poor beachfront millionaires get Atlantis" (Jackson 2001, A-23).

Jackson noted, "The vigorous multitudes who ran and biked along the Charles River in tank tops, the smiling families who picked up Christmas trees in T-shirts, and the shoppers who walked in their shorts as if this were Tampa would rather have this than blustery zero-degree wind chill" (Jackson 2001, A-23). Canada by late 2001 was in the midst of seventeen straight warmer-than-usual seasons.

Weather is never a one-way street, however. To be climatically fair, the next winter in eastern Canada (as well as the eastern United States) was colder and snowier than usual, following a global-warming tease comprising several seasons of unusual warmth. In 2003 and 2004, the northeastern quadrant of the United States and the adjacent area in Canada experienced one of their coldest winters on record. Boston (at 20.7 degrees F) and New York City (at 24.8 degrees F) had their coldest months (both January, in this case) in seventy years. At the same time, the rest of the United States was warmer than usual. Early in 2005, the Boston area was hit by a monumental blizzard that blasted easy certainties about global warming with as much as three feet of snow.

RECORD HIGHS WORLDWIDE DURING SUMMER 2003

In Phoenix, Arizona, on July 14, 2003, the high temperature reached 116 degrees F; the diurnal low the same day was 96. Windshields popped in the heat and flip–flops stuck to the pavement. In Milan, Italy, at about the same time, an African sirocco blew northward from the Sahara, making it "a little dangerous to wear stilettos [spike high-heeled shoes]; they sank half an inch into the melting tar, impaling one's legs to the spot where one's torso pitched forward, head first, like a stone being

In a haze of dust, children and elephants swim in New Delhi's Yamuna River, June 12, 2003, during a heat wave that killed more than 1,500 people across India. © RAVEENDRAN/AFP/Getty Images.

launched from a sling-shot" (Thurman 2003, 78). One day in June 2003, the temperature hit 40 degrees C (about 104 degrees F) in Milan. As stilettos were sinking into asphalt in Milan and windshields were popping in Phoenix, much of southern China was experiencing weeks of searing heat, causing shortages of drinking water and massive damage to crops. Hardest-hit Hunan, Jiangxi, Fujian, and Zhejiang provinces experienced temperatures above 40 degrees C day after day.

On August 6, 2003, London experienced its highest temperature in recorded history (since reliable records began about 1770), at 95.7 degrees F (34 degrees C). Four days later, the temperature at Heathrow International Airport peaked at 100.2 degrees F—the first three-digit reading in the climatic history of the British Isles. In Gravesend, Kent, the same day, the temperature reached 100.8 degrees F. British commuter railroads were slowed because of fear that tracks might buckle in the heat. England was hardly alone; during the same week, one location in Germany recorded an all-time record high of 40.8 degrees C (105.4 degrees F). A reporting station in Switzerland reached 41.5 degrees C (106.7 degrees F), and one in Portugal soared to 47.3 degrees C (117.1 degrees F) (McCarthy 2003, 3). The same week, some areas in southwestern Spain hit 117 degrees F. ("Europe's Heat" 2003, 2).

The year 2003 turned out to be the warmest in the entire Central England Temperature Record, which dates to 1659. The year's final average temperature of 10.65 degrees C beat the previous records of 1990 (10.63 degrees C) and 1999 (10.62 degrees C). The trend was not restricted to Europe; Albuquerque, New Mexico, for example, set a record-high minimum in 2003, at 4 degrees F above long-term averages (Fleck 2003, B-1). Power supplies failed in many urban areas, and sun-seeking tourists from the likes of Ireland and Scandinavia returned home from holidays in Italy and Greece spent in hotel rooms, with temperature outside near 100 degrees F making it too hot to sunbathe.

In France, during the same summer, an estimated 15,000 people died, including many elderly people who expired in apartments with no air conditioning. During the two weeks when the heat was most intense in France, the nation's mortality rate increased by 54 percent (Schar and Jendritzky 2004, 559). A politically charged debate enveloped the country as temperatures exceeded 100 degrees F several times during August 2003. The nation's health minister resigned. Morgues overflowed with bodies, and a quarter of France's fifty-eight nuclear power plants shut down at the height of the heat wave because rivers had become too warm to cool them properly. Paris recorded an overnight low of 78 one morning, its highest on record. During the same period, more than 1,300 people died of heat-related causes in Portugal. About 7,000 heat-related deaths were recorded in Germany, and nearly 4,200 died in both Spain and Italy. More than 2,000 people died in Britain ("Summer Heat Wave" 2003).

One analysis put Europe's death toll at 35,000 or more during the scorching summer of 2003. The Washington, D.C.–based Earth Policy Institute warned that such deaths are likely to increase as "even more extreme weather events lie ahead" ("Summer Heat Wave" 2003). Europe's heat wave of summer 2003 was the climatic crown jewel of the hottest decade in the history of the area, which reaches back at least to about 1500 CE. The average temperature was 2 degrees C (3.6 degrees F) higher than the long-term average.

"Taking into account the uncertainties in our reconstructions, it appears that the summer of 2003 was very likely warmer than any summer back to 1500," said Jurg Luterbacher, who led a 500-year study of European temperatures. Results from regional climate model simulations suggested that about every second summer will be as hot or even hotter than 2003 by the end of the twenty-first century (Luterbacher et al. 2004, 1502). The summer of 2003 in Europe was by far the hottest single season in the 500-year record examined in this study. Luterbacher and colleagues stated with a 95 percent degree of confidence that "European climate is very likely warmer than that of any time during the past 500 years" (p. 1499).

In this study, described in *Science*, Luterbacher and colleagues used temperature data from sources such as ice cores and tree rings to reconstruct the climatic history of Europe during the preceding 500 years. Against this background, the data from the past century, and particularly from its final decades, stand out as extraordinary and point strongly to human-induced global warming as the key influence (Henderson 2004, 10). While the summer 2003 heat wave in Europe was a statistical anomaly according to present average temperatures, a team of Swiss scientists found that, given expected increases in greenhouse gas levels, such events appear increasingly likely in the future (Schar et al. 2004, 332–336).

COULD EUROPE'S HEAT OF 2003 BECOME TYPICAL?

According to British scientists, Europe's scorching heat wave of 2003 will be considered typical seasonal weather by the middle of the twenty-first century and might be below average in a century. The

British team made a case that the summer of 2003 was the hottest in southern, western, and central Europe in at least five centuries. From the eastern Atlantic to the Black Sea, the mercury was 2.3 degrees C (4.14 degrees F) above average. According to their models, by the 2040s at least one European summer in two will be hotter than in 2003. "By the end of this century, 2003 would be classed as an anomalously cold summer relative to the new climate," the scientists wrote in the December 2, 2004, issue of *Nature* ("Phew" 2004). According to their models, summers in 2100 will average about 6 degrees C (10.8 degrees F) hotter than 2003's averages. "We estimate it is very likely (confidence level more than 90 per cent) that human influence has at least doubled the risk of a heat wave exceeding this threshold magnitude," wrote Peter A. Stott and colleagues in *Nature* (Stott, Stone, and Allen 2004, 610). They continued: "It seems likely that past human influence has more than doubled the risk of European summer temperature as hot as 2003, and with the likelihood of such events projected to increase 100-fold over the next four decades, it is difficult to avoid the conclusion that potentially dangerous anthropogenic interference in the climate system is already underway" (p. 613).

Separately, scientists at the French meteorological agency Meteo France said that they expect summer temperatures in France to rise by between 4 and 7 degrees C (7.2–12.6 degrees F) by 2100. "By the end of the century, a summer with temperatures as we had in 2003 will be considered a cool summer," said researcher Michel Deque ("Phew" 2004).

Across the Atlantic, as Europe sizzled, the forces of climatic denial were at work. U.S. President George W. Bush formed a "century club" composed of people willing to jog with him at his Crawford, Texas, ranch when the temperature exceeded 100 degrees F. After exercising, the joggers then reclined in Bush's air-conditioned ranch house—a comfort not available to most of the Europeans who had died in the heat. Skeptics of global warming asserted that the French needed more air conditioning, as environmentalists replied that they hadn't heretofore required mechanically cooled air to survive. Air conditioning requires copious amounts of energy, often generated by fossil fuels whose use will eventually aggravate warming.

Lake Revel, in southwestern France, which was formerly used as a reservoir for agriculture, as the French government struggled to cope with a heat wave during August 2003 that killed tens of thousands of people across Europe. © Lionel Bonaventure/AFP/Getty Images.

Heat and drought during the summer of 2003 reduced grain harvests across Europe, forcing food prices to rise worldwide. During late August 2003, the International Grains Council, an intergovernmental body, reduced its forecast of the world's grain harvest for 2003 by 36 million tons as a result of "heat and drought, particularly in Europe" (Lean 2003, 4). The damage was most severe in Eastern Europe, which harvested its worst wheat crop in three decades. In the Ukraine, the harvest fell from 21 million tons in 2002 to 5 million in 2003, while Romania has its worst crop on record. Germany was the worst hit European Union country; some farmers in the southeast regions of that country lost half their grain harvest (p. 4). World grain reserves fell sharply, continuing a four-year pattern.

Elsewhere in the world, long-term evidence of warming was plentiful. According to the Mongolian Ministry of Nature and Environment, temperatures there have risen during sixty years (1942 to 2002) by 1.56 degrees C annually, including 3.61 degrees C during winter. Rising temperatures have come with intensifying drought in many areas ("Climate Warms" 2003).

During 2001, Washington, D.C.'s annual Cherry Blossom Parade was moved up a week to coincide with earlier blooming dates in the area. (In 2003, however, after an unusually cold and snowy winter, the blossoms emerged several days later than usual.) Summer 2002 (June, July, and August) was the hottest on record in the United States except for the summers of 1934 and 1936. On September 9, 2002, the high reached 98 degrees F in Burlington, Vermont, and 97 degrees F in Bangor, Maine, both unusually hot for September. As late as October 1, 2002, Toronto, Ontario, basked in a record high of 29.2 degrees C, which broke the old daily record of 26.9 degrees C set in 1988. Beaches were open for swimming along Lake Ontario. The following winter was unusually cold and snowy in some of the same areas, however.

TWENTIETH-CENTURY ANTHROPOGENIC WARMING

Human activity has raised Earth's surface temperature significantly during the last 130 years, according to work by Robert Kaufmann of Boston University's Center for Energy and Environmental Studies and David Stern of the Australian National University's Centre for Resource and Environmental Study. A team led by Kaufmann analyzed historical data for greenhouse gas levels, human sulfur emissions, and variations in solar activity between 1865 and 1990. The greenhouse gases studied included carbon dioxide, methane, nitrous oxide, and chlorofluorocarbons 11 and 12. Using the statistical technique of "co-integration," Kaufmann and Stern compared these factors over time with global surface temperature in both the Northern and Southern hemispheres. According to its authors, this was the first study to make a statistically meaningful link between human activity and temperature independent of climate models ("Century" 2002).

This study found that the impact of human activity has been different in the two hemispheres. In the Northern Hemisphere, the warming effect of greenhouse gases was offset to some degree by the cooling effect of sulfur emissions, making the temperature effects more difficult to detect. In the Southern Hemisphere, where human sulfur emissions are lower, the effects were easier to detect. "The countervailing effects of greenhouse gases and sulfur emissions undercut comments by climate change skeptics, who argue that the rapid increase in atmospheric concentrations of greenhouse gases between the end of World War II and the early 1970s had little effect on temperature," said Kaufmann ("Century" 2002). During this period, Kaufmann said, "The warming effect of greenhouse gases was hidden by a simultaneous increase in sulfur emissions. But, since then, sulfur emissions have slowed, due to laws aimed at reducing acid rain, and this has allowed the warming effects of greenhouse gases to become more apparent" ("Century" 2002).

Peter A. Stott and colleagues used a "comparison of observations with simulations of a coupled ocean-atmosphere model" to estimate the roles of various natural and anthropogenic forcings to account for rising temperatures during the twentieth century. They found both natural and human-induced factors in the warming trend. "Natural forcings were relatively more important in the early-century warming and anthropogenic forcings have played a dominant role in warming observed in recent decades," this study observed (Stott et al. 2000, 2136). The researchers reported "the most comprehensive simulation of twentieth-century climate to date" (Zwiers and Weaver 2000, 2081). However, some of the data used in the model (such as those for the role of the sun's role) are available for only the last half of the century.

Removing the masking effects of volcanic eruptions and El Niño events from the global mean temperature record has revealed a more gradual but stronger global-warming trend over the last century, according to an analysis by Tom Wigley, a climate expert at the National Center for Atmospheric Research. This analysis supports scientists' assertions that human activity is influencing the earth's climate. The findings were published in the December 15, 2000, issue of *Geophysical Research Letters* (Wigley 2000, 4101). "Once the volcanic and El Niño influences have been removed," said Wigley, "the overall record is more consistent with our current knowledge, which suggests that both natural and anthropogenic influences on climate are important and that anthropogenic influences have become more substantial in recent decades" ("New Study" 2000).

WORLDWIDE CLIMATE LINKAGES

Rising temperatures in one area might affect vital climate patterns (including rainfall) in others, hundreds or thousands of miles away. Warming of the Indian Ocean, for example, might be at least partially responsible for declining rainfall in the Sahel region of the southern Sahara, which has contributed to devastating droughts. A modeling study by Alessandra Giannini of the International Institute for Climate Prediction in Palisades, New York, has linked rising Indian Ocean sea-surface temperatures with generally declining rainfall across the Sahel between 1930 and 2000. The Indian Ocean has warmed more rapidly than any other oceanic body on Earth (Giannini and Chang 2003, 1027–1030; Kerr 2003, 210). Studies by Giannini and colleagues link variability of rainfall in the Sahel between 1930 and 2000 to "response of the African summer monsoon to oceanic forcing, amplified by land-atmosphere interaction" (p. 1027).

Warmer oceanic temperatures tend to weaken convergence that determines the intensity of the monsoon from Senegal to Ethiopia. Another modeling study, by Mojib Latif of the University of Kiel in Germany, produced similar conclusions. "We found," he said, "that the Indian Ocean is probably the most important agent in driving decadal changes in Sahel rainfall" (Kerr 2003, 210). Rising ocean temperatures do not explain all of the Sahel's drought problems, however; Giannini and colleagues wrote that temperature change might be responsible for only 25 to 35 percent of the observed change (Zeng 2003, 1000).

INSURANCE LOSSES AND CLIMATE CHANGE

Financial companies have been addressing global warming and other aspects of human-generated climate change. Many of them are re-insurers whose businesses are already feeling the economic impact of rising, weather-related insurance claims. During July 2004, *Insurance Day* reported that the reinsurer Benfield's Hazard Research Centre had published a report stating that prospects for mitigating climate change looked "increasingly bleak," suggesting that perhaps the world already had waited too long to combat this growing threat. "The signs are that we have passed the critical point beyond which we cannot prevent significant warming nor avoid its increasingly hazardous consequences," said the report's author, Bill McGuire, Benfield professor of geohazards ("Is it Already" 2004).

Swiss Re, a large reinsurance company, has warned that the increase in the number of extreme weather events is evidence that climate change has accelerated at a more rapid pace than previously anticipated. It expects that climate change will lead to higher insurance losses in coming years. In a March 2004 report, Swiss Re said that exposure to certain extreme weather events might increase during the twenty-first century, as growing evidence indicates that human factors are responsible for a major part of the rise in global temperatures.

For the first time, Swiss Re included a section that focused on climate change in its annual report on natural catastrophes and man-made disasters. Swiss Re climate-change expert Pamela Heck told *Insurance Day* that the economic cost of global warming threatens to double every ten years, leaving insurers with a bill much larger than that of the September 11, 2001, World Trade Center bombings' loss of about $30 billion to $40 billion ("Climate Change Acceleration" 2004, 1). During 2004, insurance loses attributed to natural disasters worldwide topped $100 billion, boosted by four major hurricanes in Florida and several large typhoons in Japan. These loses did not include the massive earthquake and tsunamis that convulsed South Asia during the last week of 2004.

The Swiss Re report pointed out that 2003 was the hottest summer on record for many European countries, following extreme flash floods in some of the same areas during 2002. "We have seen the accumulation of extreme events and we can predict that there will be more extreme events in warmer climates," said Heck. "More loss years like 2002 and 2003 are likely, where there is flooding followed by heat wave[s]" ("Climate Change Acceleration" 2004, 1).

Andrew Dlugolecki, a climate-change specialist with CGNU, the world's sixth-largest insurance company, said during climate talks at the Hague that with world economic growth (measured by gross domestic product) averaging 3 percent a year and losses from climate disasters rising 10 percent a year, the two curves will cross in 2065, portending bankruptcy for the world (Brown November 24, 2000, 21; McCarthy 2000, 6). Munich Re, one of the world's largest insurance corporations, said that between 1950 and 1959 global economic losses from natural disasters—about 90 percent due to severe weather—cost the United States $42 billion. In the 1960s, costs jumped to $75.5 billion, then to $138.4 billion during the 1970s. During the 1980s, costs reached $213.9 billion. In the 1990s, these costs exceeded $660 billion (Hume 2003, A-13).

Climate change is contributing to natural disasters that the financial services industry must address, warned a group of the world's biggest banks, insurers and reinsurers during the fall of 2002. This group estimated the cost of financial losses from such events at $150 billion over the next ten years, according to "Climate Change and the Financial Services Industry," a report supported by 295 banks and insurance and investment companies that was released at the Swiss Re Greenhouse Gas Conference in Zurich. The report asserted that losses as a result of natural disasters appear to be doubling every decade and had reached $1 trillion total during the previous fifteen years. "The increasing frequency of severe climatic events, threatening the social stability or coupled with significant social costs, has the potential to stress insurers, re-insurers and banks to the point of impaired viability or even insolvency," the report concluded ("Climate-Related" 2002).

A 90 PERCENT CHANCE OF "POTENTIALLY CATASTROPHIC" GLOBAL WARMING

The Earth faces a 90 percent probability of experiencing potentially catastrophic global warming of 1.7 to 4.9 degrees C provoked by anthropogenic greenhouse gases by the end of the twenty-first century, according to a study that attempted to quantify probabilities of such a change. These probabilities were calculated by Tom Wigley of the National Center for Atmospheric Research in Boulder, Colorado, and Sarah Raper of the University of East Anglia in Norwich, Great Britain. Wigley and Raper used computer models to build on the work of the third report of the Intergovernmental Panel on Climate Change, which predicted a global temperature rise between 1.4 degrees C and 5.8 degrees C (Connor 2001, 15). The study by Wigley and Raper used more than 110,000 computations of five key variables: how much carbon dioxide is emitted, how sensitive the climate is to carbon dioxide, how much heat the world's oceans can absorb, uncertainties in the way carbon dioxide builds up, and the cooling effects of some other pollutants.

As global greenhouse gas emissions increase with the passage of time, the global window for escaping such a catastrophe narrows. The world has only fifteen years to start cutting greenhouse gas emissions if it is to stand a chance of curbing global warming, according to Bert Metz, chairman of the Intergovernmental Panel on Climate Change's mitigation committee. He told an international climate summit in Bonn during July 2001 that even a five-year delay could push the goal of stabilizing carbon dioxide in the atmosphere "beyond reach.... The first steps to fight global warming have to be taken imminently if its effects are to be contained," Metz noted (Henderson 2001).

ENDURANCE AND INTENSITY OF HEAT WAVES

Gerald A. Meehl and Claudia Tebaldi studied heat waves in Chicago during 1995 and Paris during 2003, and they used a global coupled climate model to forecast that "future heat waves in these areas will become more intense, more frequent, and longer-lasting in the second half of the twenty-first century" (Meehl and Tebaldi 2004, 994). They anticipate that areas suffering severe heat waves will probably experience more intense heat than surrounding areas: "The model show[s] that present-day heat waves over Europe and North America coincide with a specific atmospheric circulation pattern that is intensified by ongoing increases in greenhouse gases" (p. 994). They concluded, "Areas already experiencing strong heat waves (e.g., Southwest, Midwest, and Southeast United States and the Mediterranean region) could experience even more intense heat waves in the future" (p. 997).

Heat waves often are associated with semistationary high pressure at the surface and aloft, which produces clear skies, light winds, warm-air advection, and prolonged hot conditions. Meehl and Tebaldi's model suggests that these conditions will occur more frequently with increasing concentrations of greenhouse gases in the atmosphere (Meehl and Tebaldi 2004, 996).

7 DROUGHT AND DELUGE: CHANGES IN THE HYDROLOGICAL CYCLE

INTRODUCTION: WARMER AND WILDER WEATHER

While models of a warming climate generally agree that atmospheric moisture increases with temperature, theory as well as an increasing number of daily weather reports strongly indicates that changes in precipitation patterns might vary wildly across time and space. Such changes will be highly uneven, episodic, and often nasty. Both droughts and deluges are likely to become more severe. They might even alternate in some regions. By 2000, the hydrological cycle seemed to be changing more rapidly than temperatures. With sustained warming, usually wet places often seemed to be receiving more rain than before, and dry places often were experiencing less rain and subject to more persistent droughts. Some drought-stricken regions occasionally were doused with brief deluges that ran off cracking earth. In many places, the daily weather increasingly was becoming a question of drought or deluge.

Andrew Revkin of the *New York Times* summarized the situation: "A warmer world is more likely to be a wetter one, experts warn, with more evaporation resulting in more rain, in heavy and destructive downpours. But in a troublesome twist, that world may also include more intense droughts, as the increased evaporation parches soils between occasional storms" (Revkin 2002, A-10). "In a hotter climate, your chances of being caught with either too much or too little are higher," said John M. Wallace, a professor of atmospheric sciences at the University of Washington (p. A-10).

Even as flooding rains inundated some places, the percentage of Earth's land area affected by serious drought more than doubled between the 1970s and the early 2000s, according to an analysis by the National Center for Atmospheric Research (NCAR) in Boulder, Colorado. Increasing numbers of droughts occurred over much of Europe, Asia, Canada, western and southern Africa, and eastern Australia. Rising global temperatures appeared to be a major factor, said NCAR scientist Aiguo Dai ("U.S. N.S.F." 2005; Dai, Trenberth, and Qian 2004, 1117). Dai and colleagues found that the proportion of land areas experiencing very dry conditions increased from 10 to 15 percent during the early 1970s to about 30 percent by 2002. Almost half of that change was to the result of rising temperatures rather than decreases in rainfall or snowfall (Dai, Trenberth, and Qian 2004, 1117). "These results point to increased risk of droughts as human activity contributes to global warming," said Dai ("U.S. N.S.F." 2005).

Areas affected by severe drought have expanded worldwide even as the amount of water vapor in the atmosphere increased. Worldwide average precipitation also has increased slightly. However, as Dai notes, "Surface air temperatures over global land areas have increased sharply since the 1970s." Warming increases evaporation. "Droughts and floods are extreme climate events that are likely to change more rapidly than the average climate," said Dai ("U.S. N.S.F." 2005). "The warming-induced drying has occurred over most land areas since the 1970s, with the largest effects in northern mid- and high latitudes" ("U.S. N.S.F." 2005). Precipitation in the United States has run counter to that trend, however, increasing especially between the Rocky Mountains and Mississippi River.

DROUGHT AND DELUGE: A PLETHORA
OF EXAMPLES

Extreme weather is not a novelty in much of the world, of course. In recent years, however, a drumbeat of reports indicates an accelerating trend of wild weather. During August 2002, for example, Italy's grape harvest was pummeled by hail (Townsend 2002, 15). During July 2001, a fifteen-year-old girl was killed by a rare lightning strike about 100 miles from Oslo, Norway. In mid-August 2003, the Norwegian village of Atnadalen was flooded by a thunderstorm that dumped 116 millimeters of rain in twenty-four hours, twice the village's monthly average for August. During July 2003, temperatures in Norway were 3.1 degrees C above average, the highest for July in a period dating to 1866. During the summer of 2004, as if on cue to support climate models, the usual monsoon in southern Asia brought unusual deluges that killed more than 2,000 people and left millions homeless in Bangladesh, Vietnam, China, India, and Nepal. After five years of intense drought, Southern California was battered in late December 2004 and January and February 2005 by incredible amounts of rain and snow. Los Angeles received 17 inches of rain between December 27, 2004, and January 10, 2005, a record amount for a two-week period in that area. Some areas in the Sierra Nevada Mountains reported more than 10 feet of snow during the same period. At the same time, the usually wet Pacific Northwest experienced its worst drought in almost thirty years.

Seasonal snowmelt in the Sierra Nevada region started in mid–March 2004, one of its earliest dates in almost ninety years of record keeping. Stream-gauge data of California's Merced River from 1916 to the present show a shift in the onset of the melt from midspring to late winter or early spring over the last two decades, said Dan Cayan, director of the Climate Research Division at Scripps Climate Research Division and a researcher with the United States Geological Survey. In ten of the last twenty years, the melt began on or before April 1. Philip Mote, a research scientist at the University of Washington, testified to

Congress earlier in 2004 that snow packs had declined an average of 11 percent across the West since the 1970s (Hymon 2004, B-1).

The summer of 2002 featured a number of climatic extremes, especially vis-à-vis precipitation. Excessive rain deluged Europe and Asia, swamping cities and villages and killing at least 2,000 people, while drought and heat scorched the United States' West and eastern cities. Climate skeptics argued that weather is always variable, but other observers noted that extremes seemed to be more frequent than before (Revkin 2002, A-10). Also during the summer of 2002, near the Black Sea, a large tornado and heavy rains left at least thirty-seven people dead and hundreds of vacationers stranded. During the same week, in China's southern province of Hunan, seventy people died after rains caused landslides and floods. South Korea mobilized thousands of troops after a week that saw two-fifths of the average annual total rainfall (Townsend 2002, 15). During the week of May 3–10, 2003, 562 tornadoes were reported in the United States, the largest weekly total since record keeping of this kind began during the 1950s; the month of August 2004 also set a record for tornado observations in the United States. Even months not usually noted for tornadic activity seemed to be getting more; September 2004, for example, also set a record for tornado sightings in the United States.

Examples abound of increasing extremes in precipitation. November and December 2002 and January 2003 were the driest in recorded history in the cities of Minneapolis and St. Paul, Minnesota. These followed the wettest June through October there in more than 100 years. In December 2002, Omaha, Nebraska, recorded its first month on record with no measurable precipitation. In March 2003, having endured its driest year in recorded history during 2002, Denver, Colorado, recorded thirty inches of snow in *one* storm. Snowfall on Colorado's drought-parched Front Range totaled as much as eight feet in the same storm. Fifteen months later, Denver's weather let loose again; on June 9, 2004, suburbs north and west of the city received as much as three *feet* of hail. Residents used shovels to free their cars. The summer of 2003 was unusually dry in the Pacific Northwest; during the third week in October, however, Seattle recorded its wettest day on record, with

5.02 inches of rain. On the night of July 27, 2004, Dallas, Texas, recorded a foot of rain and widespread flooding—as the western United States continued to endure its worst multiyear drought in at least 500 years.

At times, the swift passage from drought to deluge can mimic Robert Frost's legendary duality of fire and ice. During November 2003, for example, the Los Angeles area was scorched by its worst wildfires on record, driven by hot, desiccating Santa Ana winds that pushed temperatures to near 100 degrees F. Less than two weeks later, parts of the Los Angeles Basin were pounded by a foot of pearl-sized hail. The island of Hispaniola (which includes the Dominican Republic and Haiti) was seared by drought during 2003 and then drowned in floods that killed at least 2,000 people during May 2004.

Roughly half the United States was under serious drought conditions during the summer of 2002. The drought was occasionally punctuated by torrential rains, however. On September 13, 2002, for example, drought-stricken Denver was inundated by floods from a fast-moving thunderstorm that caused widespread flooding. Similar events took place south of Salt Lake City, Utah. Ten days later, a flooding cloudburst inundated similarly drought-stricken Atlanta, Georgia. On September 10, 2002, six months' worth of rain fell in a few hours in the Gard, Herault, and Vaucluse departments in the south of France, drowning at least twenty people. In the French village of Sommieres, near Nimes, a usually tiny stream exploded to a width of 300 meters, cutting off road traffic. By the summer and early fall of 2004, the U.S. East Coast, which had experienced intense drought three years earlier, was drowning in record rainfall, part of which arrived courtesy of the remains of four hurricanes that had devastated Florida.

Similar reports of an intensifying hydrological cycle have been plentiful outside the United States. For instance, India, with its annual monsoon dry season that usually alternates with heavy rains, has adapted to a drought-deluge cycle. About 90 percent of India's precipitation falls between June and September during an average year, so heavy rain in Mumbai (Bombay) during late July is hardly unusual. On July 26 and 27, 2005, however, 37.1 inches of rain fell in Mumbai during twenty-four hours, the heaviest rainfall on record for an Indian city in one

An Indian family washes clothes on the banks of the Damanganga River in Vapi, western India August 6, 2004 during monsoon rains. Troops and the air force rescued people marooned by flooding, and the nationwide death toll neared 1,000. © Sebastian D'Souza/AFP/Getty Images.

diurnal cycle. The deluge contributed to more than 1,000 deaths in and near Mumbai and surrounding Maharashtra state ("Record Rainfall" 2005, A-12). The metropolitan area of 17 million was largely closed down by the rain, as several people drowned in their cars. Mass transit and telephone services stopped. Other people were electrocuted by wires falling onto flooded streets. Tens of thousands of animals also died.

During August 2002, Prague, in the Czech Republic, was inundated by flooding rains that forced 200,000 people to evacuate—the worst flooding in 200 to 500 years, depending on who was keeping the tally. The rains followed springtime drought in the same area. Debate ranged over whether this was the result of global warming provoking more intense rainstorms or a chance natural event. On August 19, 2002, Las Vegas, Nevada, in the midst of a desert, was hit by a thunderstorm that dumped three inches of rain in ninety minutes. The storm, characterized by local record keepers as a 100-year event, produced widespread flooding. During the same month, torrential rains killed 900 people in China and another 700 in Southeast Asia, India, and Nepal. Deforestation and the spread of pavement have been cited as reasons for increasing flooding in urban areas.

DROUGHT AND DELUGE: SCIENTIFIC ISSUES

As frequent and intense as such incidents may be, reports describing wild weather are episodic, not systematic. Atmospheric scientists have advocated "creation of a database of frequency and intensity using hourly precipitation amounts" (Trenberth et al. 2003, 1213). "Atmospheric moisture amounts are generally observed to be increasing... after about 1973, prior to which reliable moisture soundings are mostly not available" (p. 1211). Annual mean precipitation amounts over the United States have been increasing at 2 percent to 5 percent per decade, with "most of the increase related to temperature and hence in atmospheric water-holding capacity.... There is clear evidence that rainfall rates have changed in the United States.... The prospect may be for fewer but more intense rainfall—or snowfall—events" (pp. 1211, 1212). Individual storms might be further enhanced by latent heat release that supplies more moisture.

Generally, higher temperatures enhance evaporation, with some compensatory cooling when water is available. Increased evaporation also intensifies drought which, to some degree, compounds itself as moisture is depleted, leading "to increased risk of heat waves and wildfires in association with such droughts; because once the soil moisture is depleted then all the heating goes into raising temperatures and wilting plants" (Trenberth et al. 2003, 1212).

In extratropical mountain areas, wrote Kevin Trenberth, a scientist with the National Center for Atmospheric Research, and colleagues:

The winter snow pack forms a vital resource, not only for skiers, but also as a freshwater resource in the spring and summer as the snow melts. Yet warming makes for a shorter snow season with more precipitation falling as rain rather than snow, earlier snowmelt of the snow that does exist, and greater evaporation and ablation. These factors all contribute to diminished snow pack. In the summer of 2002, in the western parts of the United States, exceptionally low snow pack and subsequent low soil moisture likely contributed substantially to the widespread intense drought because of the importance of recycling [in the hydrological cycle]. Could this be a sign of the future? (Trenberth et al. 2003, 1212)

Scientists tend to distrust anecdotal stories, seeking, instead, consistent evidence that strongly supports a given idea or hypothesis. Thus, while the hydrological cycle seems to be changing, precipitation measurements that support such an idea have been difficult to assemble on a worldwide basis. "In many parts of the world," according to one scientific source, "we still cannot reliably measure true precipitation . . . due to rain gauge undercatch . . . precipitation is believed to be underestimated by 10 to 15 per cent" (Potter and Colman 2003, 144). Ground-level precipitation is only rarely measured over the oceans that cover two-thirds of the Earth, and satellite estimates do not measure local variability. Thus, proof of global changes in the hydrological cycle is elusive.

Despite problems of measurement, evidence of increasing variability in rain and snow has been increasing. Government scientists have measured a rise in downpour-style storms in the United States during the last century. In addition, snow levels are rising in many mountain ranges. During the last fifty years, said the University of Washington's John M. Wallace, winter precipitation in California's Sierra Nevada area has been falling more and more in the form of rain, increasing flood risks, instead of as snow, which supplies farmers and city dwellers alike as it melts in the spring (Revkin 2002, A-10). Atmospheric moisture increases more rapidly than temperature; over the United States and Europe, atmospheric moisture increased 10 to 20 percent from 1980 to 2000. "That's why you see the impact of global warming mostly in intense storms and flooding like we have seen in Europe," Trenberth told London's *Financial Times* (Cookson and Griffith 2002, 6).

While they agreed that "anthropogenic changes in atmospheric composition are expected to cause climate changes, especially enhancement of the hydrological cycle, leading to enhanced flood risk," a team of scientists wrote in *Nature* in 2003 that flooding rains in Europe had not, to date, exceeded natural variations (i.e., "a clear increase in flood occurrence rate"), despite the acute flooding of summer 2002 (Mudelsee et al. 2003, 166). Perhaps more notable than Europe's floods themselves was the fact that they were followed, the next summer, by withering drought and record heat —a definite signature of the deluge-to-drought cycle. Such evidence occurs in winter as well as summer. In late January and early February 2004, for example, Omaha received its usual seasonal snowfall (about 30 inches) in less than two weeks, in the midst of a multiple-year drought. At the same time, some areas of western Nebraska saw nary a flurry.

P.C.D. Milly, writing in *Nature* about an increasing risk of floods in a changing climate, said: "We find that the frequency of great floods increased substantially during the twentieth century. The recent emergence of a statistically significant positive trend in risk of great floods is consistent with results from the climate model, and the model suggests that the trend will continue" (Milly et al. 2002, 514–515). An increasing risk of flooding in Britain and northern Europe during the twenty-first century was quantified in the January 31, 2002, issue of *Nature* by Tim Palmer, of the European Centre for Medium-Range Weather Forecasts in Reading, England, and Jouni Raisanen of the Swedish Meteorological and Hydrological Institute in Norrkoping, Sweden.

The team analyzed the forecasts of nineteen climate models to produce an "ensemble forecast." This study also revealed that the probability of very wet summers in the Asian monsoon region would rise (Highfield 2002, 8). This was the first time that such a probability forecast of weather extremes caused by climate change had been assembled, although its methodology has been used for weather and seasonal forecasts. "Our results suggest that the probability of such extreme precipitation events is already on the increase," said Palmer. But it was "extremely difficult to verify a small increase in a statistic about extreme seasonal weather, especially over a small area like the

U.K." (p. 8). Palmer and Ralsanen wrote: "We estimate that the probability of total boreal winter precipitation exceeding two standard deviations above normal will increase by a factor of five over parts of the United Kingdom over the next 100 years. We find similar increases in probability for the Asian monsoon region in boreal summer, with implications for flooding in Bangladesh" (Palmer and Ralsanen 2002, 512).

According to a report released during 2002 by the World Water Council, which is based in France, "The economic toll of floods, droughts and other weather-related disasters has increased almost tenfold in the last four decades, a devastating pattern that must be halted with more aggressive efforts to mitigate the damage. The same report said that increasingly rapid and extreme climate changes point to a future of intensified natural disasters that will result in more human and economic misery in many parts of the world unless action is taken" (Greenaway 2003, A-5). "Most countries aren't ready to deal adequately with the severe natural disasters that we get now, a situation that will become much worse," said William Cosgrove, vice president of the council (Gardiner 2003).

The World Water Council report compiled statistics indicating that between 1971 and 1995, floods affected more than 1.5 billion people worldwide, or 100 million people a year. An estimated 318,000 were killed and more than 18 million left homeless. The economic costs of these disasters rose to an estimated $300 billion in the 1990s from about $35 billion in the 1960s (Greenaway 2003, A-5). Global warming is causing changes in weather patterns, and as growing numbers of people migrate to vulnerable areas, costs of individual weather events increase, said Cosgrove. "The forecast is that it's going to continue to get worse unless we start to take actions to mitigate global warming," he said (Gardiner 2003). Scientists cited by the World Water Council expect that climate changes during the twenty-first century will lead to shorter and more intense rainy seasons in some areas, as well as longer, more intense droughts in others, endangering some crops and species and causing a drop in global food production (Gardiner 2003).

Severe summer floods in Europe during 2002 might be an indicator of an emerging pattern, according to Jens H. Christensen and Ole B.

Christensen, who modeled precipitation patterns in Europe under warming conditions of a type that might be prominent in the area by 2070 to 2100. "Our results," they wrote in *Nature*, "indicate that episodes of severe flooding may become more frequent, despite a general trend toward drier summer conditions" (Christensen and Christensen 2003, 805). Deluges are not consistent, however; in 2003, Europe experienced heat and drought of a generational order. Farmers whose crops drowned in 2002 watched them wither and die in 2003. As much as 80 percent of the grain crop died in eastern Germany, site of some of 2002's worst floods ("Drought, Excessive Heat" 2003, 3-A). In other words, the trend toward drought or deluge will intensify as warming accelerates the hydrological cycle. A warming atmosphere will contain more water vapor, which will provide "further potential for latent-heat release during the buildup of low-pressure systems, thereby possibly both intensifying the systems and making more water available for precipitation" (Christensen and Christensen 2003, 805).

In a similar vein, Wilhelm May and colleagues expect that "in Southern Europe, the reduction of precipitation throughout the year will cause serious problems for the water supply in general and for agriculture in particular, which is very vulnerable due to its dependency on irrigation. The elongation of droughts in this region will worsen the living conditions, in particular in summer in combination with the warmer temperatures.... The dryness in Southern Europe is accompanied by an intensification of heavy precipitation events" (May, Voss, and Roeckner 2002, 27). May and colleagues also anticipate more events with heavy precipitation further north in Europe, such as in Scandinavia, where precipitation amounts are expected to increase as climate warms.

Thomas Karl, director of the National Climatic Data Center in the U.S. National Oceanic and Atmospheric Administration, said, "It is likely that the frequency of heavy and extreme precipitation events has increased as global temperatures have risen." This, he said "is particularly evident in areas where precipitation has increased, primarily in the mid- and high latitudes of the Northern Hemisphere" (Hume 2003, A-13). Studies at the Goddard Institute for Space Studies and Columbia University indicate that the frequency of heavy downpours has indeed increased and suggest that the trend will intensify. In the U.S. corn belt, for example, the average number of extreme precipitation events is predicted to jump by 30 percent over the next thirty years and by 65 percent over the next century.

DROUGHT AND DELUGE IN THE WESTERN UNITED STATES

Pushed by early warmth, the springs of 2002, 2003, and 2004 on the Great Plains exploded very quickly, compared even with the season's usual frantic pace. Tulips opened one day, and died the next. Peonies reached for the sky so quickly one could almost see their green fingers unfold. Near McCook, Nebraska, farmers said that because those years lacked adequate rainfall during the spring, wheat without irrigation would die. As if to accentuate that possibility, on the evening of April 16, 2002, the hot air was chased across a 200-mile-wide swath of central and western Nebraska by a black cloud of dust driven by sixty-mile-an-hour winds. The dust cloud was compared to the dust bowl days of the 1930s, as it reduced visibility to near zero and caused several traffic accidents. On May 22, another series of dust storms in central Nebraska provoked a ten-vehicle, chain-reaction accident that killed two people. With visibility down to twenty feet in the dust, rescue workers wore masks and goggles.

The Colorado River by 2004 was carrying only half as much water compared with the dust bowl drought of the 1930s. Utah's Lake Powell—the second biggest man-made lake in the United States—had lost nearly 60 percent of its water by 2004. Assuming present trends, the lake could lose its ability to generate electricity in three years' time (Lean 2004, 20).

Tree-ring studies of remains from the last 1,200 years indicate that the "medieval warm period," a time of unusual warmth in parts of the world, was punctuated by droughts that were longer and more intense than the one that afflicted the western part of the United States between 1999 and 2004. "Whether increased warmth in the future is due to natural variables or greenhouse [gases], it doesn't matter," said Edward

R. Cook of Columbia University's Lamont-Doherty Earth Observatory (Cook et al. 2004, 1015–1018). "If the world continues to warm, one has to worry we could be going into a period of increased drought in the western U.S. I'm not predicting that, [but] the data suggests that we need to be concerned about this" (Boxall 2004, A-17). "Compared to the earlier 'mega-droughts' that are reconstructed to have occurred around AD 936, 1034, 1150 and 1253, the current drought does not stand out as an extreme event because it has not yet lasted nearly as long," the authors wrote. "This is a disquieting result because future droughts in the West of similar duration to those seen prior to AD 1300 would be disastrous," said Cook (Boxall 2004, A-17; Cook et al. 2004, 1015–1018). "If we are just at the beginning of dramatic warming . . . we can simply expect larger, more severe fires," said Grant A. Meyer, a coauthor of the study published in *Nature* ("New Research Links" 2004, 11-A).

At ground level, the drought in the western United States occasionally was punctuated by localized deluges. On July 6, 2002, near Ogallala, Nebraska, as much as 10 inches of rain cascaded onto an area that was being plagued by extreme drought, running off the hardened soil, washing out sections of Interstate 80, killing a truck driver, and provoking evacuation of residents. Both approaches of a bridge over the South Platte River were washed out. "People I've talked to have never seen anything like this," said Leonard Johnson, mayor of Ogallala (Olson 2002, A-1). The rainfall was two to three times the amount that previously had fallen in the area during the entire year of 2002. Nearly a year later, during the night of June 22, 2003, a stagnant supercell dumped 12 to 15 inches of rain (half the area's annual average) south and east of Grand Island, Nebraska, an area that also was suffering intense drought at the time. The same storm spawned several tornadoes, killing one person and injuring several others. This storm, which destroyed large parts of Aurora, Nebraska, produced hail that was among the largest ever reported in the United States as well as a tornado that stood virtually in one place for half an hour, devastating the town of Deshler.

Irrigation water might play out just as intensifying drought arrives on the Great Plains. From the Ogallala aquifer, which supplies water to farms and ranches from Nebraska to Texas, 20 billion gallons of water more each *day* is being removed than is being replenished (Ayres 1999, 100). People in the area are just now realizing that the water that nourishes their way of life is a finite, and rapidly diminishing, resource. This knowledge has been helped along in recent years by intense drought that has rivaled the worst years of the 1930s dust bowl.

Writing in the *Handbook of Weather, Climate, and Water: Atmospheric Chemistry, Hydrology, and Societal Impacts* (2003), hydrologist Donald A. Wilhite said: "Projected changes in climate because of increased concentrations of carbon dioxide and other atmospheric trace gases suggest a possible increase in the frequency and intensity of severe drought in the Great Plains region. In a region where the incidence of drought is already high, any increase in drought frequency will place even greater pressure on the region's already limited water supplies" (Wilhite 2003, 756).

GLOBAL WARMING AND HURRICANES

The relationship (or lack thereof) between hurricane intensity and warming atmospheric temperatures is complicated by the fact that water temperatures (like air temperatures) sometimes vary over periods of several decades, with the long-term trend "signal" provoked by greenhouse gas levels. For example, water temperatures in the Atlantic Ocean, which produces nearly all the hurricanes that have an impact on the United States, have been rising steadily since the 1970s, paralleling a general global rise in air temperatures. Frequency and intensity of hurricanes (as well as the number hitting U.S. coastlines and inflicting major damage) also have been rising during the same period. Any study that takes the record back to the 1970s indicates a very tight relationship between ocean warming, hurricane intensity, and air temperatures. However, during the 1950s and 1960s, air temperatures were generally cooler than during the 1970s, but water temperatures and hurricane intensity were higher—again, on an average. By 2005, this divergence was fueling a testy debate between some hurricane experts regarding whether, and to what degree, hurricane intensity and frequency were related to the overall warming trend. This debate often spilled over into the public realm, such as when Florida and surrounding areas were smacked by four major hurricanes in 2004 and the 2005 hurricane season set records for the number of named storms in July.

A study published in the August 4, 2005, issue of *Nature* (Emanuel 2005, 686–688) indicated that the "dissipation of power" of Atlantic hurricanes had more than doubled in the previous thirty years, including a dramatic spike since 1995, with global warming and other variations in ocean temperatures working together. The study, by Massachusetts Institute of Technology climatologist Kerry Emanuel, was the first to indicate a statistical relationship between warming and storm intensity (Merzer 2005). The trend reflects longer storm lifetimes and greater intensities, both of which Emanuel associated with increasing sea-surface temperatures. "The large upswing in the last decade is unprecedented and probably reflects the effect of global warming. My results suggest that future warming might lead to an upward trend in tropical cyclone destructive potential and—taking into account an increasing coastal population—a substantial increase in hurricane-related losses in the 21st century," Emanuel wrote (Emanuel 2005, 686).

Hurricane Frances approaches Florida, September 3, 2004. Courtesy of NASA.

During the summer of 2004, Florida and adjacent areas were hit by four major hurricanes (Charley, Frances, Ivan, and Jeanne) within six weeks, as speculation mounted regarding the storms' possible relationship with global warming. Each of these hurricanes ranked in the top ten such storms to hit the United States in terms of insurance losses.

While tracing any specific storm to warming is a tenuous exercise, hurricanes might intensify generally as oceans warm. Hurricanes are essentially heat engines, so storms that approach their upper limits of intensity are expected to be slightly stronger—and produce more rainfall—in a warmer climate because of the higher sea-surface temperatures. According to a simulation study by a group of scientists at the National Oceanic and Atmospheric Administration's Geophysical Fluid Dynamics Laboratory, a 5 to 12 percent increase in wind speeds for the strongest hurricanes (typhoons) in the northwest tropical Pacific is projected if tropical sea surfaces warm by a little more than 2 degrees C. Although such an increase in the upper-limit intensity of hurricanes as the result of global warming was suggested on theoretical grounds by Emanuel more than a decade ago, this investigation is the first to examine the question using a hurricane prediction model that is being employed operationally to simulate realistic hurricane structures. The models also indicate increases in precipitation of as much as 20 percent (Knutson et al. 1998, 1018; Knutson et al. 2001, 2458).

Another study reached similar conclusions. By the 2080s, warmer seas could cause an average hurricane to intensify about an extra half step on the Saffir-Simpson scale, according to a study conducted on supercomputers at NOAA's Geophysical Fluid Dynamics Laboratory in Princeton, New Jersey. The same study anticipated that rainfall up to sixty miles from a hurricane's core could be nearly 20 percent heavier. This study is significant because it used six computer simulations of global climate devised by separate groups at institutions around the world.

Thomas R. Knutson and Robert E. Tuleya's models indicate that given sea-surface temperature increases of 0.8 to 2.4 degrees C, hurricanes would become 14 percent more intense (based on central pressure), with a 6 percent increase in maximum wind speeds and an 18 percent rise in average precipitation rates within 100 kilometers of storm centers. Tuleya is a hurricane expert who recently retired after thirty-one years at the fluid dynamics laboratory and a teacher at Old Dominion University in Norfolk, Virginia; Knutson works at Princeton University. "One implication of the results," they wrote, "is that if the frequency of tropical cyclones remains the same over the coming century, a greenhouse-gas induced warming might lead to a gradually increasing risk in the occurrence of highly destructive category-5 storms" (Knutson and Tuleya 2004, 3477).

This study of hurricanes and warming was "by far and away the most comprehensive effort" to assess the question using powerful computer simulations, said Emanuel, who had seen the paper but did not work on it. About the link between the warming of tropical oceans and storm intensity, he said, "This clinches the issue" (Revkin 2004, A-20). The study added that rising sea levels caused by global warming would lead to more flooding from hurricanes. With almost every combination of greenhouse-warmed oceans and atmosphere and formulas for storm dynamics, the results were the same: more powerful storms and more rainfall, said Tuleya (p. A-20).

While warming ocean waters are expected to intensify hurricanes and typhoons generally, other factors also are at work. For example, warming also has been associated with frequency and intensity of El Niño episodes that, at least in the Atlantic Ocean, seem to increase wind shear that inhibits development of hurricanes.

WARMING AND SPREADING DESERTS

During the first twenty years of the twenty-first century, about 60 million people are expected to leave the Sahel, a region of northern Africa that borders the fringe of the Sahara Desert, if desertification is not halted, United Nations Secretary-General Kofi Annan said on June 17, 2002, the day set aside each year by the United Nations as the World Day to Combat Desertification and Drought. In northeast Asia, "dust and sandstorms have buried human settlements and forced schools and airports to shut down," Annan said, "while in the Americas, dry spells and sandstorms have alarmed farmers and raised the specter of another Dust Bowl, reminiscent of the 1930s." In southern Europe, "lands once green and rich in vegetation are barren and brown," he said ("Global Climate Shift" 2002).

Australian government researcher Dr. Leon Rotstayn has compiled evidence indicating that air pollution is a likely contributor to catastrophic drought in the Sahel. Sulfate aerosols, tiny atmospheric particles, have contributed to a global climate shift, he said. "The Sahelian drought may be due to a combination of natural variability and atmospheric aerosol. Cleaner air in future will mean greater rainfall in this region" ("Global Climate Shift" 2002).

"Global climate change is not solely being caused by rising levels of greenhouse gases. Atmospheric pollution is also having an effect," said Rotstayn, who is affiliated with the Commonwealth Scientific and Industrial Research Organisation (CSIRO), the Australian government's climate-change research agency. Using global climate simulations, Rotstayn found that sulfate aerosols, which are concentrated mainly in the Northern Hemisphere, make cloud droplets smaller. This makes clouds brighter and longer lasting, so they reflect more sunlight into space, cooling the Earth's surface below ("Global Climate Shift" 2002). As a result, the tropical rain belt, which migrates northward and southward with the seasonal movement of the sun, is weakened in the Northern Hemisphere and does not move as far north ("Global Climate Shift" 2002). This change has had a major impact on the Sahel, which has experienced devastating drought since the 1960s. Rainfall was 20 to 49 percent lower than in the first half of the twentieth century, causing widespread famine and death ("Global Climate Shift" 2002).

GLOBAL WARMING'S IMPACT ON DISASTER RELIEF

International disaster aid will not be able to keep up with the impact of global warming, the Red Cross has said, as it reported a sharp increase in weather-related disasters during the late 1990s. In its annual

World Disasters Report, the International Federation of Red Cross and Red Crescent Societies said that floods, storms, landslides, and droughts, numbering about 200 a year before 1996, rose to 392 in 2000. Recurrent disasters "are sweeping away development gains and calling into question the possibility of recovery," the report said (Capella 2001, 15). Roger Bracke, Red Cross director of disaster relief operations, blamed the trend on global warming. As he said, "It is probable that these kind of disasters will increase even more spectacularly [in the future]" (p. 15).

Floods have accounted for more than two-thirds of the average 211 million people

A woman refugee and child in an emergency shelter during floods in Somalia. Courtesy of World Food Programme/Tom Haskell.

a year affected by natural disasters during the 1990s, according to the Red Cross. Famine caused by drought was responsible for about 42 percent of the deaths caused by natural disasters, according to this report, which said that the poor are most vulnerable to disasters; 88 percent of those affected and two-thirds of those killed by such disasters in the past ten years lived in the most impoverished countries (Capella 2001, 15).

The same report was critical of the international charity infrastructure, asserting that emergency international aid to the poorest countries declined during the late 1990s as the amount sent elsewhere rose. The report complained that many donors focus on high-profile projects to rebuild infrastructure, not people's daily livelihoods. A large amount of aid money also often is spent on studies rather than actual street-level aid. For example, the report said that nearly two-thirds of the funds spent on a flood-action plan in Bangladesh from 1990 through 1995 left the country to pay foreign aid consultants, "thereby undermining the local economy" (Capella 2001, 15).

Anticipated climatic changes related to global warming probably will deepen the gap between the richest and poorest nations, especially in parts of Africa, South Asia, and South America, according to the first worldwide assessment of warming's anticipated effects on food production. Nations in tropical climates, including India, Brazil, and much of sub-Saharan Africa, probably will experience the largest proportional losses in food production, according to this report (McFarling 2001, A-3). Most of the world's poorest peoples live in the tropics, where the effects of warming on agriculture might be the most catastrophic, with widespread starvation and malnutrition, according to the report. By contrast, people who live in cooler climates might experience gains in crop yields as higher temperatures lengthen growing seasons (p. A-3). For the poorest nations, "there is no margin for loss," said Mahendra Shah, one of the report's authors and a United Nations advisor and expert on land use from Austria's International Institute for Applied Systems Analysis (p. A-3).

Although poor nations would bear the heaviest burden as a group, some of the largest developing nations (e.g., China, Indonesia, Mexico, Chile, Congo, Kenya) would probably see increased production. Some developed countries, including Britain, the Netherlands, and Australia, could see crop yields decline as warmer, wetter weather increases diseases and pests (McFarling [July 11] 2001, A-3).

8 CLIMATE CHANGE AND THE UNITED STATES OF AMERICA AND CANADA

As in the rest of the world, climate change is leaving its fingerprints in the United States and Canada. The maple syrup industry in New England has been withering, south to north, as the region's signature winters become less severe. Water levels have been declining along with reduced snow packs in the Great Lakes watershed. Drought and warmth in the western regions of North America have provoked increasing numbers of wildfires, which add additional carbon dioxide to the atmosphere. Shrinking snow packs in the Sierra Nevada and Cascade mountains cast a pall over projections of available water for urban residents as well as agriculture. And what of the future? With additional warming, many of these problems will worsen.

CLIMATE CHANGE IN NEW ENGLAND: GOODBYE, MAPLE SYRUP

New England's maple trees require cold weather to yield the sap that can be made into maple syrup; they yield less sap in warmer winters. An analysis of syrup production between 1920 and 2000 indicated a decline in every New England state except Maine (Donn 2002). At the same time, titmice,

Sap buckets hang on maple trees in East Montpelier, Vermont, 2005. © AP/Wide World Photos.

red-bellied woodpeckers, northern cardinals, and mockingbirds are being observed more often at bird feeders in Vermont. All of these birds have migrated from more-southern latitudes as temperatures have warmed.

University of New Hampshire forester Rock Barrett, who supervised the survey, said that pervasive warming already might have doomed New England's maple syrup industry. "I think the sugar maple industry is on its way out, and there isn't much you can do about that," he said. Even in 2002, however, roughly one in four Vermont trees still was a sugar maple. Vermonters made almost 60 percent of New England's 850,000 gallons of syrup that year, according to federal farm data (Donn 2002).

A newspaper account described the industry that Barrett said is endangered:

Every year, as winter begins to melt away, Vermont's sugarhouses come back to life. They puff thick, white smoke from stainless steel evaporators that boil sap down to syrup. Punctuated by vacuum gauges, sap-carrying plastic tubing—a technology that is replacing suspended buckets—snakes through thick woods to collection tubs. Vermont kids, as always, freeze maple treats in the snow. (Donn 2002)

Much of New England could lose its maple forests during the twenty-first century in favor of the oak and hickory that are dominant further south. Already, during recent decades, most expansion in syrup production has occurred to the north, in colder Quebec province, Canada. During a decade ending in 2002, yearly production there has doubled to satisfy a booming market, which by the year 2000 surpassed the United States' output by fivefold, according to the North American Maple Syrup Council (Donn 2002). Over the last eighty years, New England's typical syrup output has dropped by more than half, from more than 1.6 million gallons a year to less than 800,000. Syrup has dwindled to a $22 million annual regional business, according to the U.S. Department of Agriculture (Donn 2002). The syrup-making season in many parts of New England has moved from March to February during recent decades, another sign that the weather is warming. "The winter weather has been interrupted, with spring . . . in between," said Toni Pease, a sugar maker in Oxford, New Hampshire, who complained of reduced sap flows from such fluctuations (Donn 2002).

The average annual temperature in Vermont and the far-northern area of New York State climbed nearly 1.6 degrees F during the twentieth century (Donn 2002). A shift away from an agrarian society and the increasing popularity of syrup substitutes also has pressured the maple syrup industry. Maine, by contrast, has cooled, slipping a yearly average of nearly a half-degree F. Meanwhile, Maine's syrup production has exploded from less than 12,000 gallons a year through most of the 1980s to 230,000 gallons by 2002. The boom is not entirely a result of the climate change; a large part of it stems from an invasion of Quebec sugar makers working Maine forests (Donn 2002). Cold winters give New England more than syrup; they discourage forest insect pests such as the maple-eating pear thrip. Researchers also say that warming promotes tree-damaging weather extremes, such as storms and drought (Donn 2002).

Maple syrup could be only one traditional casualty of rapid climate change in the northeastern regions of the United States. According to the first comprehensive U.S. federal study summarizing possible effects of global warming on the U.S. Northeast:

New England's maple trees stop producing sap. The Long Island and Cape Cod beaches shrink and shift, and disappear in places. Cases of heatstroke triple. And every 10 years or so, a winter storm floods portions of Lower Manhattan, Jersey City and Coney Island with seawater. The Northeast of recent historical memory could disappear this century, replaced by a hotter and more flood-prone region where New York could have the climate of Miami and Boston could become as sticky as Atlanta. . . . In the most optimistic projection, we still end up with a 6 to 9-degree increase in temperature," said George Hurtt, a University of New Hampshire scientist and co-author of the study on the New England region. "That's the greatest increase in temperature at any time since the last Ice Age." (Powell 2001, A-3)

Scientists who compiled this study concluded that significant warming already was taking place in New England. They noted that, on average, temperatures in the region rose 2 degrees F during the twentieth century. The scientists added that temperature rise during the twenty-first century "will be significantly larger than in the 20th century." One widely used climate model cited in the report predicted a 6 degree F increase, the other 10 degrees (Powell 2001, A-3).

Rising ocean waters present a more complicated threat. The seas around New York City have risen 15 to 18 inches in the past century, and scientists forecast that by 2050, waters could rise an additional 10 to 20 inches (Powell 2001, A-3). By 2080, storms with twenty-five-foot surges could hit New York every three or four years, inundating the Hudson River tunnels and flooding the edges of the financial district, causing billions of dollars in damage (p. A-3). "This clearly is untenable," said Klaus Jacob, a senior research scientist with Columbia University's Lamont-Doherty Earth Observatory. According to a *Washington Post* summary of the report, "Sea-level rise could reshape the entire Northeast coastline, turning the summer retreats of the Hamptons and Cape Cod into landscapes defined by dikes and houses on stilts. Should this come to pass, government would have to decide whether to allow nature to have its way, or to spend vast sums of money to replenish beaches and dunes. Complicating the issue is the fact that some wealthy coastal communities exclude nonresident tax-payers from their beaches" (p. A-3).

The dates of peak river flow in New England (from melting snow) have advanced about two weeks in the past century, according to Glenn Hodgkins, a lead researcher with the U.S. Geological Survey in Augusta, Maine. Hodgkins also has determined that the average date at which ice melted advanced by nine days in northern New England and sixteen days in southern New England between 1850 and 2000 (Sharp 2003).

HEAT-RELATED FISH KILLS ON NEW YORK LAKES

All along New York's Lake Ontario shoreline, and in many inland lakes and streams, fish died during the summer of 2002 at record rates. The deaths were attributed in part to consistently warmer temperatures. "Every day, we're getting calls from all the Finger Lakes, small ponds, a few major rivers and Lake Ontario," said Bill Abraham, the regional fisheries manager for the state Department of Environmental Conservation in Avon. "In my 30 years, I've never seen [a fish kill] so widespread" (Iven 2002). Thousands of fish had washed ashore in eastern Lake Ontario during late June and early July, several lakeside residents said.

High temperatures, which frequently climbed into the upper 80s and low 90s F, had been the main cause of fish deaths in inland bodies of water, Abraham said. According to a report in the Syracuse *Post-Standard*, "The warmer temperatures have caused an explosion in the growth of algae and rooted plants. The plants flood their water environments with oxygen during the day. During the night, however, they absorb the oxygen, choking the fish to death," Abraham explained (Iven 2002).

Overpopulation is also a factor. Most of the dead fish are alewives, also known as sawbellies or mooneyes. For the three years before 2002, Lake Ontario produced particularly large populations of these fish, which predators such as salmon and trout feed upon. The large population has thinned the supply of the alewife's preferred food, a small shrimp-like creature. Thermal upwelling also plays a role. Winds can push the warm water away from the shore, leaving cold water to replace it. The resulting drastic temperature shift can kill all fish varieties (Iven 2002).

WARMING IN THE ADIRONDACKS

During the twentieth century, the average temperature in the Adirondack mountains of Upstate New York increased more quickly than in other parts of New York state, confirming other trends that anticipate the most rapid warming at high altitudes and latitudes. From 1895 to 1999, the annual temperature in the Adirondacks rose 1.8 degrees F, while New York state as a whole warmed by 1 degree F, said climate scientist Barrett Rock, of the University of New Hampshire's Complex Systems Research Center. He asserted that the primary reason temperatures were rising so quickly was extensive logging, which removed much of the forest that had helped cool the region. "This is surprising. The Adirondacks have warmed significantly more than the rest of the state," Rock said (Capiello 2002).

The Adirondacks are considered particularly vulnerable because of their elevation and the wide variety of habitats that support species that live with in narrow temperature ranges. Very small changes in the climate could threaten those species' survival. When the climate changes, the forests of the Adirondacks could be susceptible to exotic pests and pathogens, threatening the region's timber industry and tourism economy. "Climate change is obviously affecting the Adirondacks," said Brian Houseal, the Adirondack Council's executive director. "It's going to change this entire region. When you think of forever wild, it will be for a while longer" (Capiello 2002). Rock, who used data from more than 300 federal monitoring stations nationwide to compute regional temperature rises, saw an even bigger difference between the Adirondacks and the rest of the state during winter.

Weather records indicate that between 1815 and 1950, Lake Champlain failed to freeze completely only six times. Between 1950 and 2003, however, this lake failed to freeze more than twenty-five times, yet another signal of a warming climate in this area (Lowy 2004).

WARMING AND NORTH AMERICA'S GREAT LAKES

During the winter of 2001–2002, for the first time in recorded history (and, perhaps, for long before that) North America's Great Lakes did not freeze over. The lack of persistent ice cover intensified a problem that plagued the lakes during the last decades of the twentieth century; reduced snowfall and increasing evaporation of the lakes has caused water levels to drop. Without that ice cover, millions of liters of water have been evaporating from the Great Lakes. Part of what evaporated fell as lake-effect snow at the eastern end of the lakes, creating something of a paradox: global warming, at least for the time being, was burying some lake-front areas in heavier-than-usual snows. Some of these snows were wildly variable.

Having had no snow in November 2001 and only 1.5 inches through the third week of December, residents of Buffalo, New York, greeted Christmas Eve with nearly two feet. During the ensuing week, Buffalo had another nearly three-foot storm; with a few other smaller storms, Buffalo ended the year with its snowiest month on record (about 83 inches), nearly all of which fell in one week. How is seven feet of snow in one week evidence of global warming? Lake Erie's water was much warmer than usual, and when cold air moved in suddenly, it set up the most ferocious period of lake-effect snowfall in Buffalo's recorded history.

By 2002, the Great Lakes were at their lowest point in thirty-five years, as various experts said that water levels were likely to drop even more because of unusually warm winter weather. By 2001, cargo ships were being forced to lighten their loads, and many boat ramps became inaccessible. On lakes Michigan and Huron, the water levels dropped by more than 100 centimeters, beginning in 1997, and remained 35 centimeters below average into 2002. Lake Superior was more than 15 centimeters below average, Lake Erie was 10 centimeters below, and Lake Ontario was 2.5 centimeters below its average level (Mitchell 2002). During 2003 and 2004, the pattern reversed itself, at least for a time, with snowfall in the area returning to near average.

Declining water levels on the Great Lakes are expected to play havoc with shipping. By 2050, according to one study, water levels also may fall low enough to render the hydroelectric works of Niagara Falls nearly useless. "It's going to affect everything," said Rich Thomas, chief of water management for the Army Corps of Engineers in Buffalo, where the impact of climate change has been a growing concern (Zremski 2002, A-1). Temperatures in the Great Lakes watershed are expected to rise, on average, from 3.5 to 9 degrees F during the twenty-first century. "For the Great Lakes region, the next century could bring one of the greatest environmental transformations since the end of the last Ice Age," the U.S. Environmental Protection Agency said in a study on global warming in the Great Lakes (p. A-1). Temperatures in western New York might come to resemble those of western Maryland in 2000, according to David Easterling, chief scientist for the National Climatic Data Center (p. A-1).

Jeremy Zremski, writing in the *Buffalo News*, said: "As temperatures rise, Lake Erie as we know it would be transformed. Like the rest of the Great Lakes, it would start to evaporate, meaning water levels would fall by as much as five feet over the next century. Most scientists expect the bulk of the drop to occur in the next few decades. The remaining water would be warmer and might never freeze during winter" (Zremski 2002, A-1). Fed by greater evaporation from the Great Lakes, overall precipitation in the area could increase 10 to 20 percent, according to the Environmental Protection Agency. "A lot of scientists mention getting fewer storms but of greater intensity," said Helen Domske, a researcher at the University of Buffalo's Great Lakes

Program (p. A-1). During the first few decades of warming, lake-effect snowstorms could be more frequent in Buffalo, thanks to Lake Erie's lack of ice. Easterling said, however, that temperatures eventually probably would warm to a point where lake-effect rain might become more common than snow. As a result, lake-effect snow could decrease by half within a century (p. A-1).

A comparative study of snowfall records in and outside of the Great Lakes region indicated a statistically significant increase in lake-effect snowfall in the region since the 1930s. In areas where the lake effect is small, little change has been observed. Warmer lake waters and decreased ice cover were cited by the researchers as the major reasons for the snowfall increases in lake-effect areas. A team of researchers led by Colgate University Associate Professor of Geography Adam W. Burnett published the study, "Increasing Great Lake–Effect Snowfall during the Twentieth Century: A Regional Response to Global Warming?" in the November 2003 issue of the *Journal of Climate* (Burnett et al. 2003, 3535–3542). Syracuse, New York, one of the snowiest cities in the United States, experienced four of its largest snowfalls on record during the 1990s, the warmest decade in the twentieth century. "Recent increases in the water temperature of the Great Lakes are consistent with global warming," said Burnett. "This widens the gap between water temperature and air temperature—the ideal condition for [lake-effect] snowfall" ("Global Warming Means" 2003).

The research team compared snowfall records from fifteen weather stations within the Great Lakes region with ten stations at sites outside of the region. Records dating to 1931 were examined for eight of the lake-effect areas and six of the non-lake-effect areas. Records for the rest of the sample dated back to 1950. "We found a statistically significant increase in snowfall in the lake-effect region since 1931, but no such increase in the non-lake-effect area during the same period," said Burnett. "This leads us to believe that recent increases in lake-effect snowfall are not the result of changes in regional weather disturbances" ("Global Warming Means" 2003).

Great Lakes water levels have dropped before, of course, and the low levels of the years following 1997 followed a period of relatively high water levels in the lakes, which comprise 18 percent of the world's freshwater supply. Recent low levels were notable, according to climatic experts, because of the "amount of lowering and the rapidity with which it occurred," as well as the fact that "the primary hyrdoclimatological driver was high air temperatures [increasing evaporation], not extremely low precipitation" (Assel, Quinn, and Sellinger 2004, 1150).

GLOBAL WARMING AND WILDFIRES IN WESTERN NORTH AMERICA

Fires that charred nearly three-quarters of a million acres in California during the fall of 2003 could presage increasingly severe fire danger as global warming weakens more forests through disease and drought. Warmer, windier weather and longer, drier summers could

Wildfires engulfed Summerhaven, a community on Mount Lemmon, near Tuscon, Arizona, June 20, 2003. © AP/Wide World Photos.

result in higher firefighting costs and greater loss of lives and property, according to researchers at the Lawrence Berkeley National Laboratory and the U.S. Forest Service. Both the number of out-of-control fires and the acreage burned are likely to increase, more than doubling losses in some regions, according to a study published in the scientific journal *Climatic Change*. While the study examined northern California, "the concern for Southern California would be much higher," because that region is drier for longer periods, said researcher Evan Mills of the Lawrence Berkeley lab (Thompson 2003).

According to this study, a doubling of atmospheric carbon dioxide would provoke fires that burn more intensely and spread faster in most locations. According to models developed by Jeremy S. Fried, Margaret S. Torn, and Evan Mills, the number of "escapes" (fires that exceed initial containment efforts) could double. Contained fires could also burn 50 percent more land under the warmer and windier conditions that are anticipated with a doubling of carbon dioxide concentration in the atmosphere. The researchers stressed that their projections "represent a *minimum* expected change, or best-case forecast. In addition to the increased suppression costs and economic damages, changes in fire severity of this magnitude would have widespread impacts on vegetation distribution, forest condition, and carbon storage, and greatly increase the risk to property, natural resources, and human life" (Fried, Torn, and Mills 2004, 169). The researchers projected at least a 50 percent increase in out-of-control fires in the southern San Francisco Bay area and a 125 percent increase in the Sierra Nevada foothills, with an increase of more than 40 percent in the area burned. The state's northern coast saw no significant change under the computer model and conditions used in the study (Thompson 2003).

The study's projections use conservative forecasts that did not take into account increased lightning strikes and the spread of volatile grasslands into areas now dominated by less flammable vegetation. Even potentially wetter winters simply mean more growth, providing additional fuel for summer fires, according to the study. "Fires may be hotter, move faster, and be more difficult to contain under future climate conditions," said Robert Wilkinson of the University of California at

Santa Barbara's School of Environmental Science and Management, in a federal report on the impact of climate change on California. "Extreme temperatures compound the fire risk when other conditions, such as dry fuel and wind, are present" (Thompson 2003). Damage may be aggravated by construction of homes in brushlands that are vulnerable to fires.

By midsummer 2004, as the usual wildfire season was only beginning, western North America was experiencing its worst season on record, fueled by rising temperatures and the worst drought in at least 500 years. Nearly 4 million acres of forest from Alaska to southern California had burned by the end of July. Visitors to Yosemite National Park were advised not to overexert themselves because of the "very unhealthy quality" of the air, which had been caused by pollution from fires. Usually, western wildfires reach their peak in October. In 2004, however, they started earlier than any other season on record. By early May, there had been seventy-seven fires in the state of Washington alone, compared to twenty-two by that time during 2003 (Lean 2004, 20).

SHRINKING SNOW PACK IN THE SIERRA NEVADA AND NORTHERN CALIFORNIA CASCADES

Research released during 2001 suggested that global warming would severely shrink the Sierra Nevada's snow pack, which is crucial to sustaining California's economy and population. The California water system, one of the most highly engineered networks of dams and aqueducts in the world, supports the state's 35 million people as well as some of the world's richest irrigated farmland. California's reservoirs are used for flood control as well as water supply. In the winter, they help mitigate floods; in the summer, they release water during seasonal droughts.

Snowmelt provides roughly 70 percent of the western United States' water flow. During recent decades, the snowmelt season has moved back from early spring to late winter. Spring temperatures in the Sierra Nevada region have increased 2 to 3 degrees F since 1950, bringing peak snowmelt two to three weeks earlier and prompting trees and flowers to bud one to three weeks earlier. Western rivers are seeing their peak runoff five to ten days sooner than fifty years ago.

In California, even slightly higher temperatures could lead to greater flooding, an overall decline in water supplies, a decrease in summer production of hydroelectric power, and a widespread disruption of flora and fauna in the 400-mile-long Sierra Nevada chain. Similar effects are expected in the rivers that spring from the snowfields of the Cascades in northern California (Vogel 2001, A-1). Within a lifetime, Californians could begin to see a shift in precipitation that would bring less snow and more rain to the mountains, according to scientists at the Scripps Institution of Oceanography in La Jolla, who calculated the effect of a 2 to 4 degree C temperature increase over sixty years, a pace of global warming that the Intergovernmental Panel on Climate Change considers likely.

"This whole state clings to the Sierra Nevada," said Jeffrey Mount, chairman of the Geology Department at the University of California at Davis. "The health of the Sierra Nevada, the hydrology of the Sierra Nevada, is everything to California. I don't think you can overstate that" (Vogel 2001, A-1). "The snow pack acts like a big natural reservoir," said Noah Knowles, a Scripps Institution of Oceanography researcher who used information from mountain weather stations and surveys of the state's annual snow pack to track the effect of a warming trend. "If you lose that, management-wise, it's like losing reservoir storage space" (p. A-1).

The Scripps report detailed the expected effects of a warmer, drier climate on the mountains at various elevations. Sierra Nevada mountaintops are so cold as to be immune to a slight warming of the Earth's average temperature, scientists say. But in the middle and lower elevations, a warmer atmosphere would lead to more rain and less snow. Rain runs off from these areas immediately, which would heighten the risk of floods in the Sacramento and San Joaquin valleys. Major rivers that begin in the Sierra Nevadas would also run higher in the winter, making them more dangerous (Vogel 2001, A-1).

The Scripps research on the Sierra Nevada region estimated the expected change in water content of the snow pack for each four-square-kilometer section of the mountain range. According to this analysis, the snow pack, especially at lower elevations, could shrink 80 percent or more. The *Los Angeles Times*' Nancy Vogel wrote, "Under the regimen described by the Scripps researchers, dam operators would have to keep reservoirs lower in the winter to protect against overflow during heavy rains. And there would be less chance of the reservoir refilling in the spring, because so much of the season's precipitation would have already fallen as rain, not snow" (Vogel 2001, A-1).

Researchers at the University of California at Santa Cruz came to similar conclusions. Lisa Sloan, an associate professor of earth sciences at the University of California, led a study that was published online on June 7, 2002 by *Geophysical Research Letters*. The study was described by an Enviroment News Service report: "Everybody has guessed at the effects on water resources, but now we have numbers and locations. It's a lot different from the standard arm-waving," Sloan said. "Our hope is that this kind of study will give state and regional officials a more reliable basis for planning how to cope with climate change" ("Global Warming Threatens" 2002).

This model indicated increased rainfall in northern California and unchanged precipitation in the south. Snow accumulation in the mountains would decrease in this scenario. In March, for example, the scientists' model showed an additional 8 inches of rain falling in the central Sierra Nevada area, while the depth of the snow pack at the end of March would drop by 13 feet ("Global Warming Threatens" 2002). "With less precipitation falling as snow and more as rain, plus higher temperatures creating increased demand for water, the impacts on our water storage system will be enormous," Sloan said ("Global Warming Threatens" 2002).

Elsewhere in California, a study published during the late summer of 2004 by the *Proceedings of the National Academy of Sciences* projected that by the end of the twenty-first century, rising temperatures could contribute to a sevenfold increase in heat-related deaths in Los Angeles. Given a scenario in which fossil fuel use continues at its present pace, the study forecast that summertime high temperatures could increase by 15 degrees F in some inland cities, giving them a climate much like Death Valley's in 2000. The same scenario also anticipated a reduction in the snow pack in the Sierra Nevadas of 73 percent to 90 percent, resulting in disrupted water supplies from the San Francisco Bay area to the Central Valley (Hayhoe et al. 2004, 12422). In Los Angeles, the study's worst case forecast that the number of days of extreme heat could increase by four to eight times. It projected that heat-related deaths in Los Angeles, which it said averaged 165 annually during the 1990s, "could double or triple under the moderate scenario and grow as much as seven times under the harsher one" (Murphy 2004).

WARMING AND NORTH AMERICA'S WATER SUPPLIES

The impact of warming on the mountains of California reflects similar changes in other areas of North America that rely on snow pack for water and power. A temperature rise of 2 degrees F could have dramatic impacts on water resources across western North America, according to scientific teams that have warned of reduced snow packs and more intense flooding as temperatures rise. This research was the first time that global climate modelers have worked with teams running detailed regional models of snowfall, rain, and stream flows to predict what warming will do to an area. The researchers were surprised by the size of the effects generated by a small rise in temperature ("Warmer Climate" 2001). In a warmer world, warmer winters would raise the average snow level, reducing mountain snow packs, the researchers told the American Geophysical Union in San Francisco during 2001 ("Warmer Climate" 2001).

According to the scientists' models, "Huge areas of the snow pack in the Sierra [Nevada] went down to 15 percent of today's values," said Michael Dettinger, a research hydrologist at the Scripps Institution of Oceanography in La Jolla, California. "That caught everyone's attention" ("Warmer Climate" 2001). The researchers also anticipated that by the middle of the twenty-first century, melting snow might cause streams to reach their annual peak flow as much as a month earlier than at present. With rains melting snow or drenching already-saturated ground, the risk of extreme late-winter and early-spring floods would rise, even as the diminishing snow pack's ability to provide water later in the summer would decline. Thus, consumers of water could face a frequent paradox: spring floods followed by summer drought ("Warmer Climate" 2001).

Because reservoirs cannot be filled until the risk of flooding is past, the models anticipated that within a half-century, the reservoirs would trap only 70 to 85 percent as much runoff as today. This is a particular problem for California, where agriculture, industry, a growing population, and environmental needs already compete for limited water supplies ("Warmer Climate" 2001). Observations support the models. Iris Stewart, a climate researcher at the University of California at San Diego, has found that during the last fifty years, runoff in the western United States and Canada has been peaking progressively earlier because of a region-wide trend toward warmer winters and springs ("Warmer Climate" 2001).

Water supplies in the western United States could decline by as much as 30 percent by 2050, according to one estimate. "This is just one study where we didn't find anything good: It's a train wreck," said marine physicist Tim Barnett of the Scripps Institution of Oceanography (Vergano 2002, 9-D). The Accelerated Climate Prediction Initiative pilot study late in 2002 included snow and rain forecasts for specific regions during the next five decades. Funded largely by the U.S. Department of Energy, the projections said that:

- Reduced rainfall and mountain snow runoff may reduce water released by the Colorado River to cities such as Phoenix and Los Angeles by 17 percent and cut hydroelectric power from dams along the river by 40 percent.

- Along the Columbia River system in the Pacific Northwest, water levels may drop so low that simultaneous use for irrigation and power generation will not permit any salmon spawning. Snow packs that supply the river may drop 30 percent, moving the peak runoff time forward one month.

- In California's Central Valley, "It will be impossible to meet current water system performance levels," which could hurt water supplies, reduce hydropower generation, and cause dramatic increases in saltiness in the Sacramento Delta and San Francisco Bay. (Vergano 2002, 9-D)

"The physics are very simple: Higher temperatures mean there is more rain than snow, and the spring melt comes earlier," said Barnett, who headed the two-year project (Vergano 2002, 9-D). Scientists at Scripps, the Pacific Northwest National Laboratory, the National Center for Atmospheric Research, and the University of Washington contributed to the study. Effects of climate change on water resources in the western United States were explored in depth in a special issue of *Climatic Change* on that subject, published early in 2004 (Pennell and Barnett 2004).

SNOW PACK EROSION IN THE PACIFIC NORTHWEST

What harm could global warming do in Seattle, Washington, a place where people joke that summer might come on a Saturday, if they are lucky? When Mark Twain visited, he said that the mildest winter he ever spent was a summer on Puget Sound. Seattle residents should not hold out hope that global warming will improve their chilly, soggy climate, because its most obvious evidence probably will not arrive there in the summer. Residents of the Pacific Northwest probably will feel the brunt of climate change in the winter, when snow levels will rise and wash away the next summer's irrigation and power-generating snow pack.

To forecast the severity of snow pack loss from global warming in Washington's Cascade Mountains, scientists first took a step back in time. They examined a half-century of temperature and snowfall data at weather stations from Arizona to British Columbia. "The results were striking; I was shocked by the magnitude of the (snow-moisture) declines," said University of Washington climatologist Phil Mote, who conducted much of the survey. "In some places, particularly in Oregon, we saw declines of 100 percent. It had gotten so warm there was no snow left in April at all" (Welch 2004). Mote studied snow pack records for April 1 at 145 sites in four Northwest states and British Columbia from 1950 through 1992. He found that the amount of snow between 3,000 and 9,000 feet elevation had decreased by an average of 20 percent or more. "I was surprised by the result," said Mote, who works with a group of University of Washington scientists called the Climate Impacts Group. "There's already a clearer regional signal of warming in the mountains than we expected" (Gordon 2003).

"If you think the water fights we have now are intense . . . you ain't seen nothing yet," University of Washington Professor Ed Miles said during the 2004 annual meeting of the American Association for the Advancement of Science in Seattle (Welch 2004). Miles presented evidence that moisture in snow that nourishes the West's network of rivers (and, thus, its farms and cities) has been steadily declining since at least World War II. The hardest-hit region has been the Cascades, where battles to provide enough water for fish, agriculture, and power have been intensifying for years (Welch 2004). Miles' team found that a small rise in global temperatures since 1950 already had reduced mountain snows across 75 percent of the West. In coastal regions, such as the Cascades and parts of northern California, where winter temperatures are milder, warming during the same period reduced moisture in spring snows by more than 30 percent.

Having surveyed the records, the scientists forecast future snow-pack losses using various global-warming models. Their results indicated that water content of Cascade area snows would drop by 59 percent in the coming half-century, according to their most conservative models. The likelihood of precipitation coming as rain rather than snow would be higher, and with storage reservoirs full, that would mean more early-winter flooding. Meanwhile, less snow accumulating in the mountains would mean that spring runoff probably would arrive a month to six weeks earlier than at present (Welch 2004). "The parts of the West best situated to handle water storage are the least vulnerable to temperature changes," Mote said. "Those with the least capacity to store water—the

Cascades and Northern California—are most vulnerable. So it's sort of a double-whammy" (Welch 2004).

A report by forty-six scientists from Pacific Northwest institutes and universities released in 2004 summarized existing information on future warming's probable effects for the region (available at http://inr .Oregonstate.edu). The report said that temperatures had risen 1 to 3 degrees F in the area over a century, and could rise another 3 degrees F by 2030, and 5.5 degrees F by 2050. The economic infrastructure of the region would be significantly affected by rising snow levels that would decrease the amount of snowmelt available during the summer dry season, as sea levels could rise an average of 1/2 to 1 inch per year.

WHITE CHRISTMASES SOON TO BE A MEMORY?

Statistics provided by researchers at the U.S. Department of Energy's Oak Ridge National Laboratory in Oak Ridge, Tennessee who examined weather records of sixteen cities, mainly in the northern United States after 1960, indicated that the number of white Christmases declined between the 1960s and the 1990s. In Chicago, for example, the number of white Christmases (defined as at least one inch of snow on the ground) dropped from seven in the 1960s to two during the 1990s. In New York City, the number declined from five in the 1960s to one during the 1990s, Detroit had just three white Christmases during the 1990s compared to nine in the 1960s ("White Christmases" 2001). The snowfall analysis was performed by Dale Kaiser, a meteorologist with the Carbon Dioxide Information Analysis Center at the Oak Ridge lab, and Kevin Birdwell, a meteorologist in the lab's Computational Science and Engineering Division.

WARMING-RELATED STRESSES IN CANADA

More very hot days would increase air-pollution problems in Canada's big cities, with children at greatest risk, according to a Canadian Institute of Child Health study of global warming's possible effects. Childhood asthma and respiratory problems would probably become worse if global-warming projections are accurate, according to the report. "Scientists predict that over the next century, Canadian children will live in a country that is 1.5 to 5 degrees Celsius warmer in southern Canada and as much as 5 to 7 degrees warmer in the north," asserted Don Houston, a spokesman for the institute (Bueckert 2001, A-2). "Longer, more intense summer heat waves would result in more serious smog-and-ozone episodes," the report said. It continued: "Ground-level ozone damages the cells that line the respiratory tract, causing irritation, burning, and breathing difficulties including chest tightness and pain on inhalation" (p. A-2).

Global warming could wreak havoc with Canada's prized freshwater supply over the next 100 years, sapping some of the country's hydro-electric power potential, lowering lake levels, and playing a role in more severe droughts, according to a report covered in the Toronto *Globe and Mail*. The report, "Climate Change Impacts and Adaptation: A Canadian Perspective," from Natural Resources Canada, described potential problems that could result if global surface-air temperatures increase between 1.4 and 5.8 degrees C by 2100, as projected by the Intergovernmental Panel on Climate Change (Chase 2002, A-1).

The report indicated that potential effects of climate change might include:

- Increased likelihood of severe drought on the Canadian prairies, parts of which were suffering through their second or third consecutive year of drought as the report was released;
- Ships stranded at docks and in harbors because of lowered lake or river levels, notably in the Great Lakes;
- A shrinking supply of potable water and more illness from contaminated water;
- Ruined fish habitats and spawning areas and possible loss of species;
- More financial pain for farmers caused by losses in production;
- Complete evaporation of some lakes in Arctic and sub-Arctic regions (Chase 2002, A-1).

The same report said that warming temperatures could shrink the supply of fresh water generated by melting snow during the summer, when water is in high demand. "Across southern Canada, annual mean stream flow has decreased significantly over the last 30 to 50 years, with the greatest decrease during August and September," the report said (Chase 2002, A-1). Lower water levels tend to lead to higher pollutant concentrations, whereas high-flow events and flooding increase turbidity and the flushing of contaminants into the water system. Hydroelectric power in southern Canada, where most of the population lives, also could be significantly affected by a warmer climate, the report said. It continued: "Studies suggest that the potential for hydroelectric generation will likely rise in northern regions and decrease in the south, due to projected changes in annual runoff volume" (p. A-1).

The same study estimated that excessive heat could kill more than 800 residents in the Toronto and Niagara Falls region per year by 2080, a fortyfold increase over the present-day death toll. The number of days with temperatures above 30 degrees C could double to thirty each summer by the 2030s. Ground-level ozone, a lung-damaging component in smog, was projected to double by 2080 across the region (Bueckert 2002). In addition, the report anticipated that the frequency of extreme weather events such as heat waves, windstorms, and rainstorms might increase, with associated increases in injuries, illnesses, and deaths. The incidence of water-borne diseases could rise in communities that depend on wells or in cities where sewer and storm-water drainage systems are combined, because heavy rains would increase the risk of drinking-water contamination.

Warmer weather sometimes produces more violent thunderstorms, including hail, as some Canadian apple farmers can attest. Okanagan apple grower Allan Patton said that hail used to fall about once every decade in his part of British Columbia. By 2003, however, his orchard had been pounded by hail in seven of the past ten years (Constantineau 2003, D-3). "We have to have beautiful perfect fruit to get anything for it in the marketplace," he said. "When you work all year and in a matter of 90 seconds, all your work is wiped out, it's very upsetting. It's tough to take. Very tough" (p. D-3). Patton said global warming has increased the intensity and frequency of thunderstorms at a time when farmers' crop-insurance premiums have doubled. "It's not just hail. There are bizarre wind storms where trees are blown down and a greenhouse will be lifted up but the one next to it isn't," he said (p. D-3).

BUTTERFLIES MOVE NORTHWARD ON THE CANADIAN PRAIRIES

The Edith's checkerspot butterfly has been migrating northward through the Canadian prairies, as scientists who track its movements contend that the species has become North America's first climate-change refugee. Camille Parmesan, a University of Texas biologist and lead author of *Observed Impacts of Global Climate Change in the United States*, published in November 2004 by the Pew Center on Global Climate Change, said, "The butterfly's plight is particularly poignant because it includes three endangered subspecies along the Pacific coast that are scrambling to survive increased heat and dryness that are ruining their reproductive cycles." In search of a cooler and wet-ter habitat, the butterfly "has demonstrated a clear range shift both northward and upward in elevation in response to warming tempera-tures" (Boswell 2004, A-7). The species has extended its range 100 kilometers northward in British Columbia and Alberta and climbed an average of 130 meters from previous habitats to find suitable sites for laying eggs. At the same time, the southernmost colonies of the species—in California and northern Mexico—are disappearing at an alarming rate. "Nearly 80 percent of historical populations had become extinct in the southern part of Edith's Checkerspot's range," said Parmesan.

FREAKISH WEATHER AND WILDFIRES INCREASING IN QUEBEC

Freak weather events in Canada such the 1996 Saguenay floods and Quebec's devastating 1998 ice storm are expected to occur more fre-

Flood waters from the Saguenay River ran through Chicoutimi, Quebec, damaging homes and roads, July 23, 1996. Ten died and many more were reported missing after flooding began July 20. About 12,000 people were evacuated from the area. © Carlo Allegri/AFP/Getty.

quently as the climate warms, Quebec civil protection officials have said. Climatologist Jacinthe Lacroix said that Quebecers are likely to see long spells of dry heat in summer, followed by violent thunderstorms, showers, and hail. "This will have a huge impact on our water-drainage system, which is not designed to deal with this kind of weather," he said (Sevunts 2001, A-4).

During the summer of 2002, smoke from wildfires in Quebec, aggravated by heat, drought, and lightning, drifted as far south as Baltimore, Maryland. An editorial in the Baltimore *Sun* remarked: "Was that whiff of global warming we were breathing in on Sunday? Huge tracts of forest are burning just east of Hudson Bay in Quebec, and although Americans would normally be content to ignore a natural disaster so remote and so far away, this time it was impossible because a freakish weather pattern brought the smoke southward as far as Baltimore—and even beyond" ("Quebec's Smoky Warning" 2002, 10-A).

By the first week in July 2002, roughly 400,000 acres had burned in Quebec, twice as much as would usually be expected in an entire summer. Several villages had been abandoned, and 500,000 people had lost power. The Quebec spruce forests were burning near the tundra tree line, on poor soil, and would take centuries to regenerate. "Experts have predicted that global warming will mean a northward advance of the forest line," said the *Sun*. "In Quebec, heavier precipitation. Maybe that will come, or maybe they are wrong. What we are witnessing today, in any case, is unprecedentedly warm and dry weather afflicting a fragile—and flammable—ecosystem" ("Quebec's Smoky Warning" 2002, 10-A). The fires themselves were pumping additional carbon dioxide into the atmosphere—a compounding effect of increasing wildfires worldwide.

A report in the Montreal *Gazette* attributed the increasing number and severity of wildfires in eastern Canada to logging and hydroelectric development as well as global warming. "Whenever you clear a forest for any reason, you're not only changing the ecosystem on the ground, but you're also changing the local climate," said Rene Brunet, an Environment Canada meteorologist. Clear-cutting creates dry conditions. Gone are the trees and their leafy canopies that keep the undergrowth humid and the air moist (Fidelman 2002, A-4). At the same time, global warming adds to the problem, Brunet said. "When you create a dam, you're changing the landscape. Change the color and shape of the land, and as a result, retention of moisture and solar radiation all changes" (p. A-4).

In western Canada, an editorial in the Vancouver *Province* stated:

We are running out of water, our forests are burning and where there is smoke or smog, the air is not safe to breathe. Don't think the problem will disappear with the rains as another low snow pack winter and dry summer could be in the cards. British Columbia's dangerous drought seems to be part of an intensifying global warming, which has also produced unprecedented heat waves and fires in Europe, Australia, Africa, Russia and Asia. British Columbia's Interior is a war zone with army camps in schoolyards and water bombers and helicopters in the skies. We watch the news daily to see how far the enemy fire has advanced hoping we will be safe from invasion. And we make an unimaginable sacrifice by living with a total ban on venturing into the countryside that is punishable by fines and jail. ("Editorial" 2003, A-16)

Wildfires in British Columbia during August 2003 created more greenhouse gas than 1.3 million Canadians produce in one year, according to a carbon-tracking device created by scientists in Victoria, British Columbia. The wildfires, which consumed 250,000 hectares of land, produced 15 million tons of greenhouse gases. An average Canadian produces about eleven tons a year, and British Columbia as a whole emits about 65 million tons every year (Fong 2003, A-8).

Wildfires, drought, and pests, exacerbated by steady warming of the world's atmosphere, threaten to destroy vast tracts of forested areas in Canada and other northern regions, crippling forestry and northern tourism sectors, according to Ola Ullsten, the former prime minister of Sweden, who also has served as cochair of the World Commission on Forests (Baxter 2002, A-5).

Ullsten has proposed adoption of a worldwide forest-capital index to measure the economic and environmental value of forests in regions and countries. He said that by quantifying the health of a forest and its economic value, politicians would be able to monitor forest conditions and take appropriate actions. Richard Westwood of the Centre for Interdisciplinary Forest Research at the University of Winnipeg said that human incursion already has had an effect on forests and the repercussions of global warming are simply "piling-on" to destroy forests (Baxter 2002, A-5). He said that drought causes fires as well as stress to trees, leading to pest attacks, dead and dry forests, and even more fires. "We [in Canada] are particularly vulnerable to the predicted impacts of climate change," said Terry Duguid, chairman of the Manitoba Clean Environment Commission. "Forestry is Canada's largest natural-resource industry, producing a trade surplus close to the combined surpluses for agriculture, energy, fisheries, and mining" (p. A-5).

During the summer of 2002, a team of Canadian scientists set out for the Cypress Hills in southern Alberta to determine how an isolated forest in a Great Plains environment might fare if the climate continues to warm over the next century. Their prognosis for Cypress Hills and other similar forest oases such as the Sweet Grass Hills of Montana, Moose Mountain in Saskatchewan, and Spruce Woods in Manitoba was grim, according to a report by Ed Struzik in the Edmonton *Journal*. If a catastrophic fire does not destroy these isolated forests, these "oases of trees" probably will decline more slowly from disease, insect infestation, and arid conditions (Struzik 2003, A-1).

"We can let these island forests die," said Norman Henderson, head of the Prairie Adaptation and Research Collaborative Study Group, which spearheaded the study, "Or we can move in and manage them in an intensive way. That might mean selectively harvesting timber, creating forest fire barriers, or planting new, more drought-resistant species of trees. One way or the other, the landscapes of these places aren't going to look the same as they do now whether we do anything or not" (Struzik 2003, A-1).

The destruction of the "island forests" has been characterized as a microcosm of a general decline in boreal forests across Canada. The term "boreal" refers to forests dominated by spruce and pine, which cover vast expanses of northern Canada. In an environment that is probably going to experience further warming, several networks of university and government researchers anticipate an increase in lightning storms, drought, and catastrophic forest fires that could lay waste to large regions of valuable woodlands. "The western Canadian landscape is particularly vulnerable," said Mike Flannigan, principal investigator with the government-, aboriginal-, and industry-funded Sustainable Forest Management Network, one of several groups working on the issue (Struzik 2003, A-1).

Canadian scientists' models suggest that as global warming continues, prairie landscapes may move northward. According to Struzik's report, boreal forest would give way to aspen parkland and aspen parkland would give way to prairie grasslands. The implications for a more-fiery future have important implications not only for recreation but also for the forest industry, the safety of aboriginal communities, and human health. "These direct fire emissions are significant, especially in high-fire years," said Brian Amiro, a Canadian Forest Service scientist who has been working with Flannigan (Struzik 2003, A-1). Flannigan's group anticipates that the risk of fire in western Canada might double during the forthcoming forty to sixty years.

The Canadian forestry industry will suffer economically from forest fires caused by global warming in coming decades, according to University of Alberta ecologist David Schindler. "My guess is that with climate warming, forestry will take a big hit," he said. "Right now the rates of forest cutting generally don't consider burning rates at all." Schindler said that burning rates in the western boreal forest have doubled since the 1970s (Bueckert 2002, A-3).

Wildfires have been increasing across Canada partially because average temperatures have risen about 1.5 degrees C, Schindler said. Computer models predict that average temperatures in the region will increase another 2 degrees C in the next twenty years. "If we go up another two degrees we could see at least another doubling in the incidence of forest fire," he said (Bueckert 2002, A-3). Schindler said fire suppression in Canada was effective from the 1940s through the 1970s, followed by many years during the 1980s and 1990s in which there were several large, uncontrollable fires. "You could have the whole U.S. Air Force dropping water and you wouldn't be able to contain them," Schindler said (p. A-3).

Forest fires destroy valuable economic and ecological resources as they release additional carbon dioxide to the atmosphere. "We could potentially get ourselves into a jam where our rates of carbon dioxide loss from forestry could exceed what we're putting out as fossil fuels. While we can control our fossil fuels we can't control our forest fires. That is what we call a positive feedback; it would aggravate the effects of climate warming that we're already seeing from fossil fuels. I think the prudent thing to do is try to avoid that scenario as much as possible," said Schindler (Bueckert 2002, A-3).

9 INTERNATIONAL IMPACTS

Climate change has become headline material across much of the world, but nowhere more so than Great Britain, where the London tabloids often feast on fears of weather gone wild. The British government is acutely aware of climate change's perils, the subject of many reports that argue, for example, that sizable parts of London might be abandoned to rising seas within a century. There exists on the island a palpable feeling of climatic assault by sea and atmosphere, as parts of Dover's white cliffs crumble and some of Scotland's St. Andrews Golf Course's famous links surrender to the sea (see Chapter 13, "Sea-Level Rise"). Some gardeners speculate whether they might soon be planting palm trees, while others wonder whether a breakdown of the Gulf Stream might give them winters rivaling those of Russia.

Around the world, growing seasons have been steadily lengthening, and many areas that suffered occasional droughts and heat waves now experience them more often. The broad deserts of Australia are spreading toward the thickly populated southeastern coast, as heat waves and wildfires singe the urban area of Sydney, riding hot, dry winds from the interior that are much like Southern California's Santa Anas. Meanwhile, the Japanese are contending with rapidly rising temperatures in and near Tokyo caused by general warming as well as an intensifying urban heat-island effect.

Europe's climate extremes of recent years, including heat waves, droughts, deluges, and melting glaciers in the Alps, could become routine fare there, according to a report by the European Environment Agency. The report [available at http://reports.eea.eu.int/climate-report-2-2004/en], issued in 2004, said that rising temperatures could eliminate three-quarters of Alpine glaciers by 2050 and bring repeats of Europe's mammoth floods of 2002 and the heat wave of 2003. Global warming has been evident for years, but the problem is becoming acute, Jacqueline McGlade, executive director of the Copenhagen-based agency, told the Associated Press. "What is new is the speed of change," she said (Olsen 2004). Icemelt, for example, reduced the mass of Alpine glaciers by one-tenth in 2003 alone, the report said.

WARMING AND WILD WEATHER IN GREAT BRITAIN

In central England, the growing season has lengthened by one month since 1900, with an annual temperature increase of 1 degree C. Even before Europe's searing summer of 2003, climate change had become an important factor in English political discourse. Climate change joined the political agenda under Margaret Thatcher, who taught chemistry before becoming prime minister. A staunch right-winger on most subjects, Thatcher understood the science of climate change with an acuity shared by few other political figures. The British government is among the world's most acutely aware when it comes to global warming's potential consequences. In stark contrast to the United States, where the George W. Bush administration was doing its best to edit the problem out of public consciousness at the turn of the millennium, British officialdom sounded sharp and frequent warnings. "In recent years more and more people have accepted that climate change is happening and will affect the lives of our children and grandchildren. I fear we need to start worrying about ourselves as well," said Margaret Beckett, British Environment, Food and Rural Affairs secretary (Clover 2002, 1).

The worst storm experienced by England in a decade caused road and rail chaos across the country, killed six people, and left hundreds of millions of pounds worth of damage in its wake on October 30, 2000. Torrential rain and winds up to ninety miles per hour uprooted trees, blocked roads, and cut electricity supply lines across southern England and Wales. According to newspaper reports, a tornado ripped through a trailer park in Selsey in West Sussex less than forty-eight hours after a similar twister had devastated parts of Bognor Regis. In Yorkshire, the first blizzards of the winter coincided with flash floods. English weather record keepers said that October's rainfall in East Sussex, one of the driest parts of the country, had been nearly three times its average, at 226 millimeters (9 inches). September also had been exceptionally wet.

Marilyn McKenzie Hedger, head of the United Kingdom Climate Impacts Program based at Oxford University, said: "These events should be a wake-up call to everyone to discover how we are going to cope with climate change" (Brown [October 31] 2000, 1). Michael Meacher, U.K. environment minister, said that while it would be foolish to blame global warming every time extreme weather conditions occur, "The increasing frequency and intensity of extreme climate phenomena suggest that although global warming is certainly not the sole cause, it is very likely to be a major contributory factor" (p. 1). The storm included the lowest barometric pressure on record in the United Kingdom during October, 951 millibars.

The next day, John Prescott, deputy prime minister, said that extreme weather events must now be regarded as usual fare in Britain as global warming takes hold. Railways, power lines, and flood defenses must adapt, he said. Government officials, local authorities, and representatives of emergency services and environment agencies were summoned to a meeting in London a day later, where Prescott demanded action, saying: "Our infrastructure should be robust enough, and our preparations rigorous enough, to withstand the kind of weather we have just experienced." He continued: "We aren't putting the amount of resources and investment in for what we call more extreme conditions, which we must now accept [are] normal. We have to ask ourselves: Should our power lines come down every time we have such storms? Should 1,000 trees fall across our railway lines in the southeast? Should we do more to prevent flooding? Are our drainage systems really adequate?" (Brown [November 1] 2000, 1).

A British government report released during February 2001 warned that climate changes during the first half of the twenty-first century could cause death and destruction on a major scale in Britain unless immediate preventive actions are taken. The report, prepared by the Expert Group on Climate Change and Health, was the first official assessment of the effect of global warming on health. It asserted that rising sea levels and severe storms are likely to cause "catastrophic" flooding that could devastate tens of thousands of homes. Hotter summers are expected to result in 30,000 extra cases of skin cancer (unless pollution can be curtailed), 10,000 extra cases of food poisoning, and an extra 3,000 lives lost in heat waves. Warmer winters are expected to reduce the toll of deaths among the elderly caused by the cold, saving an estimated 20,000 lives a year (Laurance 2001, 2). The same report said that flooding of low-lying coastal areas would become more likely as a result of rising sea levels and storm surges. "Flooding that spreads inland from coastal areas may be catastrophic," it said, leaving perhaps tens of thousands of people temporarily homeless. Local National Health Service resources "could be overwhelmed" (p. 2).

During the previous thirteen years, Britain had experienced three winter storms of a severity that earlier would have been expected once every 200 years (Laurance 2001). The next year, in 2002, the British government warned that homes and businesses in the United Kingdom worth 222 billion pounds sterling in total (about $350 billion in U.S. funds) would be threatened by devastating flooding under global-warming conditions during coming decades. The report, published by the Energy Savings Trust, estimated that 5 million people living in 1.8 million homes, one in every thirteen people in the United Kingdom, risk being inundated by rising seas and increasing rainfall. By one account, the report was "the starkest official assessment yet of the human cost of climate change in Britain" (Lean 2002, 1).

The same report said that more than three-fifths of Britain's best farmland, 3.5 million acres, also would be threatened by flooding. The report was issued during September 2002, as flash floods struck across Britain "from Inverness—which was cut off, with parts of the city under 5 feet of water—to the Isle of Wight, and hitting Bognor Regis,

Swanage, Glasgow and parts of Fife and Ayrshire in between. Parts of London were also submerged, shutting down rail and Tube [subway] services, while central Europe suffered its worst-ever floods"—showing, as German Chancellor Gerhard Schroder told the Earth Summit in Johannesburg, South Africa, that "Climate change is no longer a skeptical forecast but a bitter reality" (Lean 2002, 1).

About half of the 222 billion pounds sterling in potential property damage anticipated in this report was forecast in the Thames River region around London. According to the report, the homes of 750,000 Londoners are at risk, and the capital's future "as an international centre for trade and commerce" is threatened (Lean 2002, 1). "The Thames Barrier was designed on the basis that it would not have to be closed more than 10 times a year. But in the year 2000–2001 it had to be shut 24 times, nearly as often as the total for its entire history until then" (p. 1).

For the first time, this report urged Britain's government to adopt policies that would persuade people to give up automobiles powered by internal-combustion engines. While the report admitted that much could be done to reduce emissions of carbon dioxide by improving the efficiency of cars and introducing cleaner fuels, it concluded: "A long-term policy aimed at slowing down and ultimately reducing car ownership, as well as use, will be necessary to have any real impact on transport emissions" (Lean 2002, 1).

This report was released in mid-September 2002. Within six weeks, Britain was being battered by intense storminess that was described in the *London Guardian*:

> As big storms go, yesterday morning's [October 27, 2002] was not quite on the scale of October 16 1987 when some 15 million trees were uprooted in a wild night that changed the face of southern England. But ecologists said yesterday it was a timely reminder that the terrifying weather once assumed to take place only every 250 years is now liable to occur far more frequently. The 1987 storm, which left a devastating trail across 10 counties and killed 18 people, was said to be the greatest in Britain since 1709. British Broadcasting Corporation weatherman Michael Fish had said the

night before that "no hurricane" was expected, but wind speeds of over 115 m.p.h. in Norfolk and along the south coast were not far short of those usually seen in the Caribbean and the Pacific. (Vidal [October 28] 2002, 3)

The same account continued:

What no one expected in 1987 was that less than 27 months later there would be another storm of almost the same intensity. The 1990 event battered the West Country and Wales, again blocking roads and railways for several days and cutting power supplies in some areas for up to a week. Forecasters had predicted yesterday's plunging barometric pressures but few had any idea until just hours before that wind speeds in South Wales and the West Country would rival those of 1987. Then, the strongest gust was 122 m. p. h. at Gorleston, Norfolk, with similar strengths in Hampshire and Sussex. The damage caused yesterday is less than that of 1987, partly because there are fewer mature trees to be uprooted. After two major storms in 15 years, Britain's stock of great trees is smaller and those remaining are more resilient. (Vidal [October 28] 2002, 3)

AN "ORDERLY RETREAT" OF GOVERNMENT FROM LONDON?

During 2004, a panel of sixty British climate-change experts released a government-sponsored report, "Future Flooding," which asserted that the homes of as many as 4 million Britons might be at risk of inundation by 2050. The report said that the national cost of flooding might rise from $2.6 billion a year about 2000 to $52 billion annually by 2080. Some government officials warned that the government might be forced to consider an "orderly retreat" from London because parts of the 2,000-year-old city are below sea level. Paul Samuels, who was leading a Europe-wide study of flooding, said London could be "mostly gone in the next few centuries" (Melvin 2004, 3-A).

The flooding report said that Britain must create "green corridors" in cities to act as safety valves into which floodwaters can be channeled. It said that parts of some urban areas might have to be abandoned and oil refineries moved inland. Many homes, it warned, might become un-insurable. Samuels, who suggested that the government retreat from London, was working on the premise that the tidal section of the Thames River would rise as much as to 3 feet in a century, a situation exacerbated by the subsidence of the land on which some of London is built (Melvin 2004, 3-A).

English scientists have considered different scenarios for high, medi-um, and low rates of emission of greenhouse gases, predicting the following changes in the British climate by 2080: The average tem-perature could rise from 2 to 3.5 degrees C, probably with greater warming in the south and east. Generally, the climate could become be like that of Normandy, the Loire, or Bordeaux, varying according to the level of global greenhouse gas emissions. Hot days in summers would be more frequent, with some above 40 degrees C (104 degrees F) in lowland Britain under the high-emission scenario. According to the highest-emission projections, the United Kingdom's summer rainfall might decrease by 50 percent and winter rainfall might increase by 30 percent. Snowfall would decrease throughout Britain. Scotland might experience 90 percent less snow, according to the highest-emissions scenario. Sea levels could rise by 26 to 86 centimeters (10 to 34 inches). The probability of extreme storm surges would increase from one in fifty years to nine in ten years under the high-emissions scenario (Clover 2002, 1).

After several years of excessive rains, Great Britain by 2003 was experiencing a drought. At Hyde Park, London, between February and April 2003, rainfall measured 2.9 inches. During the same period in 2002, the total precipitation had been 5.4 inches, and in 2001, con-cluding the wettest twelve months on record in England and Wales, 10.8 inches of rain fell at the same station during the same period. Commented Ross Clark in the *Daily Telegraph* (London): "The only certain thing about our weather is that it is becoming more biblical. The story of the past few months has been one of 40 days of drought, followed by a 40-day deluge, concluded by another 40-day drought" (Clark 2003, 15).

From mid-January to mid-February 2002, according to the British Meteorological Office, the average temperature was 8.2 degrees C, which was 4.4 degrees C above the long-term average (Vidal [February 23] 2002, 3). A spokesman from the same office said that while spells of mild weather had not heretofore been unknown in Britain during the winter, they now were becoming common, with January and February temperatures increasing significantly. "We have seen a definite reduction in the frequency of winter frosts over the past 20 years. It's mostly because of increased cloud cover and higher night temperatures" (p. 3). "We are predicting spring to come two and a half weeks earlier than the long-term average," said Nick Collinson of the Woodland Trust. "October was the warmest on record, and this is already the warmest January in years. On average spring now comes a week earlier than 30 years ago. There's definitely something going on" (p. 3).

Tim Sparks, a research biologist at the Centre for Ecology and Hydrology in Cambridge, said at the time that England's spring of 2002 could be its earliest in 300 years. "Winter is being shrunk at both ends. We had a very late autumn last year and most of January and February have been mild," he said. (Vidal [February 23] 2002, 3). "Nobody has heard the first cuckoo yet—usually [in] mid-April—but the first frog-spawn was observed on December 10, the first primroses as early as October and the first snowdrops a week before Christmas" (p. 3).

BRITAIN'S WARMEST OCTOBER ON RECORD

As concern about global warming was smothered in the United States by a deluge of terrorism concerns following the September 11, 2001, attacks on New York City and Washington, D.C., Britain experienced its warmest October on record. "Butterflies from America, birds from the Mediterranean, mushrooms afoot and spring flowers in bloom—all part of Britain's warmest-ever October, according to records kept since 1659," said one account (McCarthy 2001, 12). The average temperature for the month was 13.5 degrees C, much higher than the long-term average of 10.6 degrees C. The previous record for an October, 13.0 degrees C, had been set in 1969. During October 2001's warmest spell, in the middle of the month, several places, including London, reached a temperature of more than 25 degrees C.

According to a report in the *Independent* (London), "Wildlife has responded accordingly. Bees, moths and dragonflies were all to be seen yesterday [October 29, 2001] at the Wetland Centre nature reserve in Barnes, southwest London. Butterflies have been on the wing astonishingly late—meadow browns, holly blues and speckled woods have been prolific and are still visible. On October 21, Martin Warren, of the charity Butterfly Conservation, saw a Silver-studded Blue in the New Forest, the latest recorded [sighting] for the area for more than 70 years" (McCarthy 2001, 12). An influx of monarch butterflies from North America was blown by storm winds across the Atlantic when they should have been migrating to Mexico.

In parts of Devon, daffodils poked through the soil out of cycle. In the Gordano Valley, Somerset, a scientist affiliated with the charity Plantlife found flowering marsh marigolds, which are usually in bloom from March to June. Plantlife volunteers also reported seeing other spring flowers in bloom, including cow parsley and blue fleabane (McCarthy 2001, 12). Many birds that should have been migrating southward from southern Europe reversed course and flew northward to Britain. A spokesman for the British Meteorological Office said: "October really has been an astonishingly warm month. We can't say this is by itself proof of global warming, but it is certainly another piece of the jigsaw" (p. 12).

Of the record October warmth in Britain, an observer wrote in the *London Times*:

> Keats would have been as bewildered as the bees that "think warm days will never cease." Our autumn is not his: the mists have been dispersed by hot sun or drowned by tropical downpours, and mellow fruitfulness has been rejuvenated by trees in bud, flowers in bloom and grass still pushing up sturdily across a million lawns. In Kent the moss'd cottage trees may soon no

longer bend with apples, as farmers grub up their orchards and plant walnuts, sunflowers and vines more commonly found in the distant oases of Uzbekistan than the Garden of England. No longer are stubble-plains touched with rosy hue; most are now under several feet of water, as glassy floods spread far across the land. ("Walnuts" 2001)

MONSOON BRITAIN

By 2000, England's summers were becoming quasitropical. According to one newspaper report in 2002, dated August 11: Temperatures above 30 degrees C and brutal storms pounded London with several usual weeks of rain in a few minutes. Scientists believe such events are merely a taste of things to come. Experts at the Tyndall Centre of Climate Change at the University of East Anglia warn that "[precipitation is] becoming increasingly erratic as the planet heats up, offering humankind arguably its biggest challenge to date" (Townsend 2002, 15).

In mid-August, 2002, a sudden deluge dumped twenty days' worth of rain on London in thirty minutes, "causing chaos as the antiquated drainage systems and transport infrastructure failed almost instantly and flash floods turned the city's streets into rivers. Glasgow was still re-covering from its own 'freak storm' that overwhelmed parts of the sewage system. The crisis forced scores of families into temporary ac-commodation and thousands had to boil drinking water as floods contaminated the supply" (Townsend 2002, 15). On August 10, 2002, seaside towns in North Yorkshire were hit by flash floods, as more than 100 people were evacuated from their homes.

During August 2004, a deluge over London flushed 600,000 tons of sewage into the Thames. Two weeks later, mud and debris carried by another deluge swept into the British coastal town of Boscastle, which lies at the confluence of three rivers, sweeping away homes and cars. The flood was fed by more than six inches of rain (2.5 inches of which fell in one hour) and complicated by an unusually high tide that impeded the water's flow into the sea. During the same series of storms, residents of Knighton reported a shoal of fish falling from the sky during a thunderstorm. August 2004 was among England's wettest in recorded history, with copious flash floods, landslides, lightning strikes, hailstorms, and at least fourteen tornadoes.

Occasional wet summers are hardly unknown in England, however. Boscastle itself was flooded in 1847, 1957, and 1958, as well as in 2004. The frequency of

Spectators watched as high tides caused the Thames River to rise over quays, flooding automobiles at Richmond upon Thames. © Angelo Hornak/Corbis.

Britain's climatic violence is new, however. John Turnpenny, senior research adviser at the Tyndall Centre, said that he expected "monsoon" conditions to become more prevalent, particularly in southern England. That means the violent storms of the early twenty-first century could become commonplace. In addition, heat waves resembling those of 1995 and 2003—during which maximum temperatures remained above 25 degrees C for seventeen days in August—are "forecast in Britain for two out of every three years by the time that this century ends" (Townsend 2002, 15).

Elaine Jones, a British government climate specialist, said that the Victorian infrastructure of London, particularly its sewers, needs to be improved if it is to maintain its position as one of the world's great cities. "The drains can't take the water away. The result is that you risk sewage in the streets and all those associated public health risks," Jones told the *Observer* (Townsend 2002, 15).

PALM TREES AND BANANA PLANTS IN ENGLISH GARDENS?

Traditional English gardens have been changing as climate warms. As described in an Associated Press dispatch carried in Canada's *Financial Post*:

The fabled English garden with its velvety green lawn and vivid daffodils, delphiniums and bluebells is under threat from global warming, leading conservation groups said late in 2002. Within the next 50 to 80 years, palm trees, figs and oranges may find themselves more at home in Britain's hotter, drier summers, the National Trust and the Royal Horticultural Society said, releasing a new report on the impact of climate change. *Gardening in the Global Greenhouse: The Impacts of Climate Change on Gardens in the U.K.* was commissioned by the two organizations and the government, as well as water, forestry and botanical organizations. (Woods 2002, S-10)

Reading University scientists Richard Bisgrove and Paul Hadley forecast fewer frosts, earlier springs, higher year-round temperatures, increased winter rainfall (with increased risk of flooding), and hotter, drier summers (with increased risk of drought) (Woods 2002, S-10). "While there will be greater opportunities to grow exotic fruits and subtropical plants, increased winter rainfall will present difficulties for Mediterranean species which dislike waterlogging," said Andrew Colquhoun, director general of the Royal Horticultural Society (p. S-10).

A large number of cool-weather plants are likely to suffer, according to this report, including "snowdrops, crocuses, rhododendrons, ferns and mosses, along with bluebells and daffodils. It won't be impossible to grow delphiniums, the Royal Horticultural Society said, but they will be more difficult. . . . The society said gardeners can expect to see more palms, grapes, citrus fruit, figs and apricots, as well as colorful climbers like plumbago and bougainvillea. New pests from southern climates—such as the rosemary beetle, berberis sawfly and the lily beetle—are now established in Britain, the society said" (Woods 2002, S-10).

The Chelsea Flower Show in May 2002 "strongly reflected the trend for Mediterranean-style plants suitable for dry conditions" (Johnson 2002, 5). Climate models for England projected warmer, drier summers and wetter winters. Landscape architects are faced with a paradox of finding plants that can survive hotter, drier summers while building landscapes that can carry off a larger volume of winter floodwaters. Guy Barter, head of the Royal Horticultural Society advisory service, said: "Olive trees, grapes, avocados and even banana plants could all become common garden features. The air could be full of the scent of acacia. . . . We will also see more gardens with heat-resistant trees, and cacti and yucca. But the problem will be flooding in winter" (p. 5).

The May 2002 edition of the British Meteorological Society's magazine *Weather* reported evidence that the English growing season has been lengthening by an average of a day a year; in 2000 it was the longest on record, at 330 days (leaving a "winter" of 35 days). "If the trend continues, it is possible we'll have a year-round growing season within a generation," said Tim Mitchell of the Tyndall Centre (Johnson 2002, 5). The *London Guardian* reported "daffodils blooming near Buntingford in Hertfordshire in mid-February." It added, "Helped by one of mildest winters on record, other spring indicators such as budding blackthorn and the first butterfly have arrived early. Scientists say the trend will continue. The evidence of the past five weeks suggests that this has been one of the warmest starts to a year since records began more than 300 years ago. The signs that spring has not just stirred but has actually sprung up to three weeks earlier than usual are now everywhere. Daffodils have been blooming in Scotland for weeks, toads have been on their ancient migratory marches to breeding ponds well before schedule, hawthorn buds are bursting in hedgerows, and the lesser celandine and other vernal indicators are flowering in woodland" (Vidal [February 23] 2002, 3).

Biological and botanical events that heretofore had taken place in March, or even April, began in February during 2002: Lambs were "gamboling around," the elder leaves and forsythia flowers were out, and the horse chestnuts and blackthorn were budding. Tits were nesting, frogs were spawning, and gardeners were mowing lawns that had grown shaggy during the winter (Vidal [February 23] 2002, 3).

BRITAIN'S COASTAL EROSION: GOODBYE, WHITE CLIFFS OF DOVER

The 300-foot-high white cliffs of Dover are crumbling into the sea. "Last winter," by one account, "some 100,000 tons of Dover Cliffs crashed down onto the beach" (Simons 2001, 14). Erosion during the year 2000 was the worst since record keeping began. A 600-foot-long stretch of cliff front at Dover crumbled into the English Channel on January 31, 2000. At nearby Beachy Head, a 200-foot chalk formation fell to beach level on April 3. "We've had 26 major rockfalls since last May," said Steven Judd, area manager for the National Trust, the environmental organization that owns most of the Dover coast (Reid 2001, A-23). "We've never had that much activity before," Judd pointed out. "Many people have suggested that this is evidence of global warming, and so it's going to get worse" (p. A-23). "We've probably got tens of thousands of years left," said Judd. "But the cliff line will be eroding steadily back, west toward London as the chalk falls" (p. A-23).

Beach Road, Happisburgh, on the north Norfolk coast of England, "used to be a cliff-top drive with a grand view out to sea. Now the road stops short at a sign saying 'Road Closed' with a 30-foot drop on the other side" (Simons 2001, 14). The rest of the road and several adjacent homes have fallen into the sea. Great chunks of England's southernmost east coast are falling into the sea, as up to six feet of shoreline a year crumbles away. Slowly rising seas together with torrential rainfalls erode the soft clay that crumbles under the pounding of waves roiled by ocean storms of increasing violence. On the east coast, according to one English account, "more than 69,000 homes, 7,000 businesses, and large tracts of agricultural land are at risk, at a total cost estimated at more than 7 billion pounds sterling. Only vast concrete sea defenses can save these threatened coasts; otherwise it is a question of managed retreat and let nature take its course" (p. 14).

The crumbling of the soft, porous white chalk is a natural phenomenon, but the speed with which the cliffs are now disintegrating is new. Some cliff-top buildings are threatened, and some lighthouses have been moved inland. Walking trails that skirt the edges of the cliffs have been abandoned. Heavy rains destabilize Dover's trademark white chalk from the top as increasing storminess erodes the cliffs from below. The chalk that forms the cliffs was built up over 60 million years from the white shells of microscopic animals in an ancient sea. A chalk land bridge connected Britain and mainland Europe until about 10,000 years ago, when flooding created the English Channel. The channel is actually a chalk canyon, with tall white cliffs on the English coast as well as along the opposite coast of France (Reid 2001, A-23).

Southern England has been subsiding about 30 centimeters a year, compounding erosion provoked by rising seas and more intense storminess, all of which increase the effects of storm surges. One estimate anticipates that the sea level at East Anglia might rise 80 centimeters by the middle of the twenty-first century. Scotland, which is not subsiding, can expect a six-inch rise in sea levels during the next fifty years; England can expect sea levels to rise a foot during the same period (Simons 2001, 14).

CLIMATE CHANGES IN SCOTLAND

More than 12,000 forts, castles, and other archaeological sites that stand along Scotland's shorelines are being eroded into the sea "as rising tides and relentless storms, brought about by global warming, east away at the coastline" (Grant 2001, 5). Among the sites most at risk is Scrabster Castle, near Thurso, which was built in the twelfth century by the Bishop of Caithness. This structure is referenced in the Vikings' Orkney Inga Saga, which dates to 1196. Dunbar Castle, where Mary, Queen of Scots, sought refuge after the murder of her second husband, also is threatened by the sea. Tom Dawson, a researcher at St. Andrews University's School of History, said, "Parts of the coastline are receding up to one meter a year" (p. 5). A third-century fort on Orkney, the Borch of Borwick, also has been crumbling into the sea "with pottery and human skeletons falling from ruins into the water" (p. 5).

In another sign that Scotland's climate has warmed, hops, which give beer its bitter flavor, have invaded the Scottish countryside. Hops, heretofore confined to the southern British Isles and the southern English Midlands, have been sighted in large patches in at least a dozen locations along roadways near Pathhead, Midlothian. Botany specialist Brian Moffat was quoted in *Herald* (Glasgow) as saying: "Finding one hop plant here is usually a little bit of a freak event; finding so many is really extraordinary" ("Changes in Climate" 2000, 13).

Two of Scotland's five ski resorts were put up for sale early in 2004 after they suffered large financial losses. The owners of Glencoe and Glenshee put the resorts on the market after deciding that they could no longer afford to keep them open. Mild winters and lack of snow have left the winter sports industry in Scotland in very poor financial shape. With the pace of global warming accelerating, some in Scotland expected that its skiing industry could cease to exist within twenty years.

Fred Last, president of the Royal Caledonian Horticultural Society, who has recorded changes in plants for a quarter-century, said he had observed significant changes in Scotland. Based upon his continued observations of 800 species, he said: "There is good evidence that some plants are flowering earlier than before but the patterns of flowering have changed. Some flowering plants are showing about three weeks earlier, gardeners will need to think about their herbaceous borders to fill in any possible gaps that may occur" (Stewart 2002, 8). "If all the predictions are right, 2050 could turn out to be an excellent year for Scottish wine," quipped the *Herald* (Glasgow) (Crilly 2002, 11).

LONGER GROWING SEASONS IN EURASIA

Researchers at Boston University and the American Geophysical Union have described dramatic changes in the timing of both the appearance and fall of leaves (as recorded by weather satellites during the past twenty-one years), according to data published in the September 16, 2000, issue of the *Journal of Geophysical Research*. The growing season is now almost eighteen days longer in Eurasia, with spring arriving a week early and autumn delayed by ten days. In North America, the growing season appears to be as much as twelve days longer (Dube 2001). Satellite weather photos used in this research were compared with temperature data from several thousand meteorological stations in the United States and around the world.

While a longer growing season might seem like cause for celebration, researcher Ranga Myneni, an associate professor in the department of geography at Boston University, stated that the news is cause for concern. "What is good for plants is not necessarily good for the planet" (Dube 2001).

Wolfgang Lucht and colleagues documented a two-decade-long trend toward earlier spring bud-burst and increased leaf cover in northern latitudes that was briefly interrupted only following the 1991 eruption of Mt. Pinotubo during 1992 and 1993 (Lucht et al. 2002, 1687). "We conclude," they wrote, "that there has been a greening trend in the high northern latitudes, associated with the gradual lengthening of the growing season . . . and reduced snow-cover extent" (p. 1688).

WILDFIRES, DROUGHT, AND FLOODS INCREASE IN AUSTRALIA

Human-induced global warming was a key factor in the severity of the 2002 drought in Australia, the worst in the country's history, according to a report by World Wildlife Fund (WWF) Australia. The report, titled *Global Warming Contributes to Australia's Worst Drought*, by David Karoly, James Risbey, and Anna Reynolds, associated the intensity of the drought with warming temperatures. By early 2003, 71 percent of Australia was in serious or severe drought. In some areas, the drought was pervasive—97 percent of New South Wales was drought-stricken, according to the report (Macken 2003, 68). The WWF asserted that Australian businesses and its federal government continued

Drought during 2002 killed thousands of kangaroos in Sturt National Park in the far northwest of New South Wales, Australia. © John Carnemolla/Australian Picture Library/Corbis.

to equivocate over global warming. As southeast Australia experienced its worst drought in a century, hundreds of kangaroos headed toward the capital city, Canberra, turning up on golf courses and sports fields in search of fertile grass. Hundreds of them were shot to death "by professional gunmen as growing numbers [of people] perceived [the kangaroos as] a threat to the capital's 320,000-strong population" ("Why We're All" 2004).

The WWF report was part of an effort by Australian environmental organizations to convince the federal government, led by Prime Minister John Howard, to reverse its policy and sign the Kyoto Protocol. During 2002, Australia recorded its highest average March–November daytime maximum temperature on record. The temperature across the country was 1.6 degrees C higher than the long-term average and 0.8 degrees C higher than the previous record. The record heat accelerated evaporation rates, intensifying drought. The report indicated that the intensity of the 2003 fire season, which claimed five lives and destroyed more than 500 homes, also was associated with climate change. Like the drought, 2002's firestorms were described as a once-in-a-century event. "This is the first drought in Australia where the impact of human-induced global warming can be clearly observed," said Karoly, who is associated with the University of Oklahoma's School of Meteorology (Macken 2003, 68).

Like most major droughts in Australia, this one was associated with an El Niño weather pattern. The report said, however, that while higher temperatures are expected during El Niño–triggered droughts, temperatures during the 2002 drought were extraordinary when compared with the five previous major droughts in Australia since 1950, registering an average maximum temperature that was 1.2 degrees C higher than any of them. Increased temperatures led to higher levels of evaporation. This combination of factors also led to a greater danger of bush fires and dry conditions in forests that could provide potential fuel for explosive firestorms, according to the report (Macken 2003, 68).

During the summers of 2002 and 2003, wildfires pushed by raging hot and dry winds from Australia's interior seared parts of Canberra, charring hundreds of homes, killing four people, and forcing thousands to flee the area. "I have seen a lot of bush fire scenes in Australia . . . but this is by far the worst," Australia's Prime Minister Howard said ("Australia Assesses" 2003, A-4). Flames spread through undergrowth and exploded as they hit oil-filled eucalyptus trees. The 2002–2003 drought reduced Australia's gross domestic product by 1 percent and cost $6.8 billion in exports. It reduced the size of Australia's cattle herd by 5 percent and its sheep flock by 10 percent (Macken 2004, 61). Recovery was impeded during 2004 by continuing drought.

The Australia Green Party urged the country's federal government to expand an inquiry into the devastating bushfires to consider global warming. Bob Brown, a Green Party Australian senator, said that new research indicates that record daytime temperatures and unprecedented rates of water evaporation made the 2002–2003 drought the worst on record. The extreme dryness of vegetation arising from global warming was what made this fire season so devastating, he said ("Greens Want" 2003).

Even as drought aggravated the wildfires in Australia, some parts of the country were dealing with deluges. Speaking in Sydney, Tomihiro Taniguchi, vice chairman of the Intergovernmental Panel on Climate Change, said that recent Australian flooding and similar weather conditions around the world were further evidence that the impact of global warming was beginning to be felt. While he admitted it was difficult to directly link deluges in New South Wales to global warming, he added, "We can say with high confidence that the likelihood of flooding will increase in the future" (Maynard 2001, 17). The higher temperatures create greater seawater evaporation, which in turn produces more precipitation in coastal areas.

Forecasts released during May 2001 by the Commonwealth Scientific and Industrial Research Organization (CSIRO) of Australia suggested that "large areas of already marginal farmland . . . will become unusable, while fragile wetland and coral ecosystems, including the Great Barrier Reef, will be threatened. The rise in the sea level—as much as 88 centimeters over the 1990 level by 2100—also would exacerbate damage from cyclones" (Marsh 2001, 18). Average temperatures in Australia are forecast to increase by 1 to 6 degrees C by 2070. This CSIRO research said that shifts in rainfall patterns would aggravate already-arid conditions across much of Australia, with more intense cyclones in tropical regions. Warmer weather also might allow pests to spread.

Australia's Queensland state might lose half of its wet tropical highland rainforest, including many of its most rare animals, because of global warming, another Australian report asserted. The state's emblem, the koala, was at risk because of rising carbon dioxide levels, which could strip the gum leaf (the koala's principal food) of its nutritional value. The same goes for many other vulnerable species. The report, released February 4, 2002, by Climate Action Network Australia, "represents one of the most comprehensive pictures yet of the local ecological effects of global warming" (Ryan 2002, 1). More than half of Australia's eucalyptus species are unlikely to survive a 3 degree C average temperature rise, according to the report. In addition, higher carbon dioxide levels are expected to reduce carbohydrates and nitrogen in leaves, undermining food supplies for animals such as koalas (p. 1).

The study reviewed possible damage to Queenland's tropical areas, asserting that "90 Australian animal species, including a third of those on the endangered list, also [are] likely to suffer in the hotter, more extreme climate forecast this century" (Ryan 2002, 1). David Hilbert, principal research scientist at the CSIRO Tropical Forest Research Center, said even a 1 degree C temperature rise would devastate half of the rainforests of northern Queensland's wet tropics.

"C.S.I.R.O. atmospheric scientists are now predicting 2 to 5 degree C of warming by the end of the century, so a 1 degree C change is liable to happen within 30 to 50 years," Hilbert said (Ryan 2002, 1). The Murray-Darling Basin, for example, faces a 12 to 35 percent reduction in average flow by 2050 as Australia's southern regions grow hotter and drier, adding to the pressures on its overextracted and salt-affected rivers (p. 1). The Great Barrier Reef also runs a risk of heat-related coral bleaching, which affected 16 percent of the world's reefs in 1998, on top of existing threats such as land-based runoff (p. 1). In addition, Australia's alpine ecosystems are expected to shrink to six high mountaintops with continued warming, with any remaining snowfields probably disappearing within 100 years, the report asserted.

Drought might become chronic in Australia if modeling done by CSIRO proves accurate. According to Kevin Hennessey, of the agency's Atmospheric Research Climate Impact Group: "There is a consistency between our modeling and the reality of Australia's weather. Our modeling suggests Australia will become warmer and drier in the future as a result of global warming. By 2030 most of Australia will be between 0.5 and 2 degrees warmer, and potentially 10 per cent drier" (Macken 2004, 61).

Anecdotal evidence of warmth and drought in Australia is supported by scientific study. Neville Nicholls, of Australia's Bureau of Meteorology Research Centre (Melbourne), wrote in *Climatic Change*:

Rainfall over nearly all of Australia during the cooler half of the year (May–October) was well below average in 2002. Mean maximum temperatures were very high during this period, as was evaporation. This would suggest that drought conditions (precipitation minus evaporation) were worse than in previous recent periods with similarly low rainfall (1982, 1994). Mean minimum temperatures were also much higher during the 2002 drought than in the 1982 and 1994 droughts. The relatively warm temperatures in 2002 were partly the result of a continued warming evident in Australia since the middle of the twentieth century. The possibility that the enhanced greenhouse effect is increasing the severity of Australian droughts, by raising temperatures and hence increasing evaporation, even if rainfall does not decrease, needs to be considered. (Nicholls 2004, 323)

JAPAN: HEAT ISLAND TOKYO

According to the *Daily Yomiuri* of Tokyo, temperatures in several Japanese cities averaged between 3.2 and 3.9 degrees C above long-term averages during the winter of 2001–2002. The increase in temperatures was most notable in and around Tokyo. During the early-morning hours (midnight to 5 a.m.), average temperatures in Tokyo have risen by 7.2 degrees F in a century. In 1900, the number of "tropical nights" with minimum temperatures above 77 degrees F was zero to five in an average summer. By the early twenty-first century, the number of such nights reached thirty to forty during most summers (Brooke 2002, A-3). On July 20, 2004, the temperature in Tokyo hit a record-breaking 39.5 degrees C (103.1 degrees F), the hottest temperature recorded there since records began in 1923.

Tokyo winters also have become milder, with nighttime temperatures rarely dropping below freezing. Snow in Tokyo is increasingly rare. None at all fell there during the winter of 2001–2002. Leaves used to start turning color in the end of November, said Shinsuke Hagiwara, chief researcher of National Institute for Nature Study. "Now they only start in mid-December" (Brooke 2002, A-3). During the spring of 2002, cherry blossoms in Tokyo opened so early that when Prime Minister Junichiro Koizumi held the government's annual cherry blossom viewing party in April, the blossoms had fallen from the trees. A type of mosquito carrying dengue fever, usually found in warmer places, by 2002 had expanded its range to sixty miles north of Tokyo, according to Mutsuo Kobayashi, a medical entomologist (p. A-3).

Rural Japan is awash in anecdotes of unusual warming as well. In rural areas as well as Tokyo, cherry blossoms have been blooming earlier than in the past, leaving people to "hold their blossom-viewing parties under leaves instead of flowers" (Hatsuhisa 2002, n.p.). The Prefecture of Niigata on the Sea of Japan, two hours north of Tokyo by bullet train, which once was known for heavy "ocean-effect" snows carried by cold air from Siberia, reported a scarcity of snow during the winter of 2001–2002, followed by heavy snow in 2005–2006. During March 2002, no snow fell there for the first time since weather records had been kept (Hatsuhisa 2002, n.p.). Many resorts that depend on snowfall (mostly for skiing) were forced to close. During the same month, two-thirds of Japan's weather-observation stations reported their highest temperatures in a statistical record that, in most cases, reaches to 1886.

REFERENCES: PART II. THE WEATHER NOW— AND IN 2100

Alley, Richard B., ed. *Abrupt Climate Change: Inevitable Surprises*. Committee on Abrupt Climate Change, Ocean Studies Board, Polar Research Board, Board on Atmospheric Sciences and Climate, Division of Earth and Life Sciences, National Research Council. Washington, D.C.: National Academy Press, 2002.

Assel, Raymond A., Frank H. Quinn, and Cynthia E. Sellinger. "Hyrdoclimatic Factors of the Recent Record Drop in Laurentian Great Lakes Water Levels." *Bulletin of the American Meteorological Society* 85 (8) (August 2004): 1143–1150.

"Australia Assesses Fire Damage in Capital." Associated Press in *Omaha World-Herald*, January 20, 2003, A-4.

Ayres, Ed. *God's Last Offer: Negotiating a Sustainable Future*. New York: Four Walls Eight Windows, 1999.

Baxter, James. "Canada's Forests at Risk of Devastation: Global Warming Could Ravage Tourism, Lumber Industry, Commission Warns." *Ottawa Citizen*, March 5, 2002, A-5.

Bleach, Stephen. "The Naked Truth." *Sunday Times* (London), August 24, 2003, Travel Section, 1.

Boswell, Randy. "Southern Butterfly's Trek North Cited as Proof of Global Warming." *Edmonton Journal*, November 19, 2004, A-7.

Boxall, Bettina. "Epic Droughts Possible, Study Says: Tree Ring Records Suggest That If Past Is Prologue, Global Warming Could Trigger Much Longer Dry Spells Than the One Now in West, Scientists Say." *Los Angeles Times*, October 8, 2004, A-17.

Brooke, James. " 'Heat Island' Tokyo Is Global Warming's Vanguard." *New York Times*, August 13, 2002, A-3.

Brown, Paul. "Global Warming: It's with Us Now; Six Dead as Storms Bring Chaos throughout the Country." *London Guardian*, October 31, 2000, 1.

———. No headline. *London Guardian*, November 1, 2000, 1.

———. "Islands in Peril Plead for Deal." *London Guardian*, November 24, 2000, 21.

Bueckert, Dennis. "Climate Change Could Bring Malaria, Dengue Fever to Southern Ontario, Says Report." Canadian Press in *Ottawa Citizen*, October 23, 2002, www.canada.com/news/story.asp?id={B019135A-4FD8-4536-A2F0-908F13560CB2.

———. "Climate Change Linked to Ill Health in Children." Canadian Press in *Montreal Gazette*, June 2, 2001, A-2.

———. "Forest Fires Taking Toll on Climate: CO₂ from Increased Burning Could Overtake Fossil Fuels as a Source of Global Warming, Prof Warns." Canadian Press in *Edmonton Journal*, September 19, 2002, A-3.

Burnett, Adam W., Matthew E. Kirby, Henry T. Mullins, and William P. Patterson. "Increasing Great Lake–Effect Snowfall during the Twentieth Century: A Regional Response to Global Warming?" *Journal of Climate* 16 (21) (November 1, 2003): 3535–3542.

Capella, Peter. "Disasters Will Outstrip Aid Effort as World Heats Up: Rich States Could Be Sued as Voluntary Assistance Falters, Red Cross Says." *London Guardian*, June 29, 2001, 15.

Capiello, Dina. "Adirondacks Climate Growing Hotter Faster." *Albany Times-Union*, September 21, 2002.

"Century of Human Impact Warms Earth's Surface." Environment News Service, January 24, 2002. http://ens-news.com/ens/jan2002/2002L-01-24-09.html.

Chambers, Kevin. "Fewer Frosty Days." Weather.com, January 17, 2001. www.weather.com/weather_center/full_story/full3.html.

"Changes in Climate Bring Hops Northward." *Herald* (Glasgow, Scotland), September 29, 2000, 13.

Chase, Steven. "Our Water Is At Risk, Climate Study Finds." *Toronto Globe and Mail*, August 13, 2002, A-1. www.globeandmail.com/servlet/ArticleNews/PEstory/TGAM/20020813/UENVIN/national/national/national_temp/6/6/23/.

Christensen, Jens H., and Ole B. Christensen. "Severe Summertime Flooding in Europe." *Nature* 421 (February 20, 2003): 805.

Clark, Ross. "Rain, Rain Come Again: The Long Dry Spell Is Making Gardeners Anxious." *London Daily Telegraph*, May 17, 2003, 15.

"Climate Change Acceleration Will Push Claims Bills Higher." *Insurance Day*, March 4, 2004, 1.

"Climate Change and the Financial Services Industry." United Nations Environment Programme Finance Initiative, No date. www.unepfi.net.

"Climate-Related Perils Could Bankrupt Insurers." Environment News Service, October 7, 2002. http://ens-news.com/ens/oct2002/2002-10-07-02.asp.

"Climate Warms Twice as Fast." British Broadcasting Corporation Monitoring Asia-Pacific, January 6, 2003. (Lexis).

Clover, Charles. "Climate Conference Pounds 1,000 Grant to Switch to 'Green' Cars." *London Daily Telegraph*, November 21, 2000, 9.

———. "2002 'Warmest for 1,000 Years.'" *London Daily Telegraph*, April 26, 2002, 1.

Connor, Steve. "Catastrophic Climate Change 90 Per Cent Certain." *London Independent*, July 20, 2001, 15.

Constantineau, Bruce. "Weather Wreaking Havoc on British Columbia Farms: Global Warming Has Increased the Intensity and Frequency of Weather Events, Making It Difficult for Farmers to Compete in the Marketplace." *Vancouver Sun*, March 3, 2003, D-3.

Cook, Edward R., Connie A. Woodhouse, C. Mark Eakin, David M. Meko, and David W. Stahle. "Long-Term Aridity Changes in the Western United States." *Science* 306 (November 5, 2004): 1015–1018.

Cookson, Clive, and Victoria Griffith. "Blame for Flooding May Be Misplaced: Climate Change Global Warming May Not Be the Reason for Recent Heavy Rainfall in Europe and Asia." *London Financial Times*, August 15, 2002, 6.

Crilly, Rob. "2050 To Be Good Year for Scottish Wine; Global Warming Will Bring Grapes North." *Glasgow Scotland, (Herald)*, November 20, 2002, 11.

Dai, Aiguo Dai, Kevin E. Trenberth, and Taotao Qian. "A Global Dataset of Palmer Drought Severity Index for 1870–2002: Relationship with Soil Moisture and Effects of Surface Warming." *Journal of Hydrometeorology* 5 (6) (December, 2004): 1117–1130.

Donn, Jeff. "New England's Brilliant Autumn Sugar Maples—and Their Syrup—Threatened by Warmth." Associated Press, September 23, 2002. (Lexis).

"Drought, Excessive Heat Ruining Harvests in Western Europe." Associated Press in *Daytona Beach (FL) News-Journal*, August 5, 2003, 3-A.

Dube, Francine. "North America's Growing Season 12 Days Longer: 'What Is Good for Plants Is Not Necessarily Good for the Planet': Expert." *Canada National Post*, September 5, 2001. www.nationalpost.com.

Editorial. *Vancouver Province*, September 8, 2003, A-16.

Emanuel, Kerry. "Increasing Destructiveness of Tropical Storms Over the Past 30 Years." *Nature* 436 (August 4, 2005): 686–688.

Enger, Tim. "If This Is Spring, How Come I'm Still Shoveling Snow?" Letter to the editor. *Edmonton Journal*, May 8, 2003, A-19.

"Europe's Heat Wave Toll Tops 19,000." Associated Press in *Omaha World-Herald*, September 26, 2003, 2.

Fidelman, Charlie. "Longer, Stronger Blazes Forecast." *Montreal Gazette*, July 11, 2002, A-4.

Fleck, John. "Dry Days, Warm Nights." *Albuquerque Journal*, December 28, 2003, B-1.

———. "Jack Frost's Nip Arrives a Bit Later." *Albuquerque Journal*, October 26, 2002, A-1.

Fong, Petti. "Greenhouse Gas Chokes Sky after Wildfires." CanWest News Service in *Calgary Herald*, September 23, 2003, A-8.

Fried, Jeremy S., Margaret S. Torn, and Evan Mills. "The Impact of Climate Change on Wildfire Severity." *Climatic Change* 64 (May 2004): 169–191.

Gaardner, Nancy. "State Enjoying 'Exceptional' Warmth." *Omaha World-Herald*, December 4, 2001, A-1, A-2.

Gardiner, Beth. "Report: Extreme Weather on the Rise, Likely to Get Worse." Associated Press Worldstream, International News, London. February 27, 2003 (Lexis).

Giannini, A., R. Saravanan, and P. Chang. "Oceanic Forcing of Sahel Rainfall on Interannual to Interdecadal Time Scales." *Science* 302 (November 7, 2003): 1027–1030.

"Global Climate Shift Feeds Spreading Deserts." Environment News Service, June 17, 2002. http://ens-news.com/ens/jun2002/2002-06-17-03.asp.

"Global Warming Means More Snow for Great Lakes Region." AScribe News-wire, November 4, 2003. (Lexis).

"Global Warming Threatens California Water Supplies." Environment News Service, June 4, 2002. http://ens-news.com/ens/jun2002/2002-06-04-09.asp#anchor3.

Gordon, Susan. "U.S. Pacific Northwest Gets Reduced Supply of Snow, Climate Study Says." *Tacoma (WA) News-Tribune*, February 7, 2003. (Lexis).

Grant, Christine. "Swelling Seas Eating Away at Country's Monuments." *The Scotsman*, December 24, 2001, 5.

"Great Lakes Water Levels Dropping: Lowest in 35 Years." Associated Press in *Canada National Post*, January 4, 2002. www.nationalpost.com.

Greenaway, Norma. "Disaster Toll from Weather Up Tenfold: Droughts, Floods Need More Damage Control, Report Says." *Edmonton Journal*, February 28, 2003, A-5.

"Greens Want Global Warming Examined in Bushfire Inquiry." Australian Associated Press, January 21, 2003. (Lexis).

"*Guardian* Special Report: Global Warming." *London Guardian*, November 14, 2000. www.guardian.co.uk/globalwarming/story/0,7369,397255,00.html.

Hansen, James, R. Ruedy, M. Sato, and K. Lo. "Global Warming Continues." *Science* 295 (January 11, 2002): 275.

Hatsuhisa, Takashima. "Climate." *Journal of Japanese Trade and Industry*, September 1, 2002. (Lexis).

Hayhoe, Katherine, Daniel Cayan, Christopher B. Field, Peter C. Frumhoff, Edwin P. Maurer, Norman L. Miller, et al. "Emissions Pathways, Climate Change, and Impacts on California." *Proceedings of the National Academy of Sciences* 101 (34) (August 24, 2004): 12422–12427.

Henderson, Mark. "Past Ten Summers Were the Hottest in 500 Years." *London Times*, March 5, 2004, 10.

———. " 'World Has 15 Years to Stop Global Warming.' " *London Times*, July 21, 2001. (Lexis).

Highfield, Roger. "Winter Floods 'Five Times More Likely.' " *London Daily Telegraph*, January 31, 2002, 8.

"How About That Weather? The Answer Is Blowing in the Wind, Rain, Snow. . . ." *Washington Post*, January 26, 2005, C-16.

Hume, Stephen. "A Risk We Can't Afford: The Summer of Fire and the Winter of the Deluge Should Prove to the Naysayers That If We Wait Too Long to React to Climate Change We'll Be in Grave Peril." *Vancouver Sun*, October 23, 2003, A-13.

Hymon, Steve. "Early Snowmelt Ignites Global Warming Worries: Scientists Have Known Rising Temperatures Could Deplete Water Sources, but Data Show It May Already Be Happening." *Los Angeles Times*, June 28, 2004, B-1.

"Is It Already Too Late to Stop Global Warming?" *Insurance Day*, July 14, 2004. (Lexis).

Iven, Chris. "Heat Strangles Fish." *Syracuse (NY) Post-Standard*, July 9, 2002. www.syracuse.com/news/poststandard/index.ssf?/base/news-0/10262073067431.xml.

Jackson, Derrick Z. "Sweltering in a Winter Wonderland." *Boston Globe*, December 5, 2001, A-23.

Jayaraman, K. S. "Monsoon Rains Start to Ease India's Drought." *Nature* 423 (June 12, 2003): 673.

Johnson, Andrew. "Climate to Bring New Gardening Revolution: Hot Summers and Wet Winters Could Kill Our Best-Loved Plants." *London Independent*, May 12, 2002, 5.

Karl, Thomas R., and Kevin E. Trenberth. "Modern Global Climate Change." *Science* 302 (December 5, 2003): 1719–1723.

Kerr, Richard A. "Warming Indian Ocean Wringing Moisture from the Sahel." *Science* 302 (October 10, 2003): 210–211.

Knutson, Thomas R., and Robert E. Tuleya. "Impact of CO_2-Induced Warming on Simulated Hurricane Intensity and Precipitation: Sensitivity to the Choice of Climate Model and Convective Parameterization." *Journal of Climate* 17 (18) (September 15, 2004): 3477–3495.

———, Robert E. Tuleya, Weixing Shen, and Isaac Ginis. "Impact of CO_2-Induced Warming on Hurricane Intensities as Simulated in a Hurricane Model with Ocean Coupling." *Journal of Climate* 14 (2001): 2458–2469.

———, Robert E. Tuleya, and Yoshio Kurihara. "Simulated Increase of Hurricane Intensities in a CO_2-Warmed Climate." *Science* 297 (February 13, 1998): 1018–1020.

Koslov, Mikhail, and Natalia G. Berlina. "Decline in Length of the Summer Season on the Kola Peninsula, Russia." *Climatic Change* 54 (September 2002): 387–398.

Laurance, Jeremy. "Climate Change to Kill Thousands, Ministers Warned." *London Independent*, February 9, 2001, 2.

Lean, Geoffrey. "Experts Prove How Warming Changes World." *London Independent*, February 18, 2001, 12.

———. "Hot Summer Sparks Global Food Crisis." *London Independent*, August 31, 2003, 4.

———. "U.K. Homes Face Huge New Threat from Floods." *London Independent*, September 15, 2002, 1.

———. "Worst U.S. Drought in 500 Years Fuels Raging California Wildfires." *London Independent*, July 25, 2004, 20.

Lederer, Edith M. "U.N. Report Says Planet in Peril." Associated Press, August 13, 2002. (Lexis).

Lowy, Joan. "Effects of Climate Warming Are Here and Now." Scripps-Howard News Service, May 5, 2004. (Lexis).

Lucht, Wolfgang, I. Colin Prentice, Ranga B. Myneni, Stephen Sitch, Pierre Friedlingstein, Wolfgang Cramer, et al. "Climatic Control of the High-Latitude Vegetation Greening Trend and Pinatubo Effect." *Science* 296 (May 31, 2002): 1687–1689.

Luterbacher, Jurg, Daniel Dietrich, Elena Xoplaki, Martin Grosjean, and Heinz Wanner. "European Seasonal and Annual Temperature Variability, Trends, and Extremes Since 1500." *Science* 303 (March 5, 2004): 1499–1503.

Macken, Julie. "The Big Dry: Bushfires Re-ignite Heated Debate on Global Warming." *Australian Financial Review*, February 17, 2003, 68.

———. "The Double-Whammy Drought." *Australian Financial Review*, May 4, 2004, 61.

Marsh, Virginia. "Australia Expected to Become Hotter." *London Financial Times*, May 9, 2001, 18.

May, Wilhelm, Reinhard Voss, and Erich Roeckner. "Changes in the Mean and Extremes of the Hydrological Cycle in Europe under Enhanced Greenhouse Gas Conditions in a Global Time-Slice Experiment." In *Climatic Change: Implications for the Hydrological Cycle and for Water Management*, ed. Martin Beniston, 1–30. Dordrecht, Germany: Kluwer Academic Publishers, 2002.

Maynard, Roger. "Climate Change Bringing More Floods to Australia." *Singapore Straits Times*, March 14, 2001, 17.

McCarthy, Michael. "Climate Change Will Bankrupt the World." *London Independent*, November 24, 2000, 6.

———. "The Four Degrees: How Europe's Hottest Summer Shows Global Warming Is Transforming Our World." *London Independent*, December 8, 2003, 3.

———. "Global Warming: Warm Spell Sees Nature Defying the Seasons: As 150 Countries Meet in Morocco to Discuss Climate Change, Britain's Natural World Responds to Record Temperatures." *London Independent*, October 30, 2001, 12.

McFarling, Usha Lee. "NASA Finds 2002 Second Warmest Year on Record." *Los Angeles Times* in *Calgary Herald*, December 12, 2002, A-5.

———. "Warmer World Will Starve Many, Report Says." *Los Angeles Times*, July 11, 2001, A-3.

Meehl, Gerald A., and Claudia Tebaldi. "More Intense, More Frequent, and Longer Lasting Heat Waves in the 21st Century." *Science* 305 (August 13, 2004): 994–997.

Melvin, Don. "There'll Always Be an England? Study of Global Warming Says Sea Is Winning." *Atlanta Journal-Constitution*, June 5, 2004, 3-A.

Merzer, Martin. "Study: Global Warming Likely Making Hurricanes Stronger." *Miami Herald*, August 1, 2005. (Lexis).

Milly, P. C. D., R. T. Wetherald, K. A. Dunne, and T. L. Delworth. "Increasing Risk of Great Floods in a Changing Climate." *Nature* 415 (January 30, 2002): 514–517.

Mitchell, John G. "Down the Drain: The Incredible Shrinking Great Lakes." *National Geographic*, September 2002, 34–51.

Moberg, Anders, Dmitry M. Sonechkin, Karin Holmgren, Nina M. Datsenko, and Wibjörn Karlén. "Highly Variable Northern Hemisphere Temperatures Reconstructed from Low- and High-Resolution Proxy Data." *Nature* 433 (February 10, 2005): 613–617.

Mudelsee, Mandred, Michael Borngen, Gerd Tetzlaff, and Uwe Grunewald. "No Upward Trends in the Occurrence of Extreme Floods in Central Europe." *Nature* 425 (September 11, 2003): 166–169.

Murphy, Dean. "Study Finds Climate Shift Threatens California." *New York Times*, August 17, 2004. www.nytimes.com/2004/08/17/national/17heat .html.

"New Research Links Global Warming to Wildfires across the West." *Los Angeles Times* in *Omaha World-Herald*, November 5, 2004, 11-A.

"New Study Shows Global Warming Trend Greater without El Niño and Volcanic Influences." Environmental Journalists' Bulletin Board, December 13, 2000, environmentaljournalists@egroups.com.

Nicholls, Neville. "The Changing Nature of Australian Droughts." *Climatic Change* 63 (2004): 323–336.

O'Driscoll, Patrick. "2005 Is Warmest Year on Record for Northern Hemisphere, Scientists Say." *USA Today*, December 16, 2005, 2-A.

Olsen, Jan M. "Europe Is Warned of Changing Climate." Associated Press Online, August 19, 2004. (Lexis).

Olson, Jeremy. "Flash Flooding Closes I-80." *Omaha World-Herald*, July 7, 2002, A-1.

Palmer, T. N., and J. Ralsanen. "Quantifying the Risk of Extreme Seasonal Precipitation Events in a Changing Climate." *Nature* 415 (January 30, 2002): 512–514.

Pennell, William, and Tim Barnett, eds. "The Effects of Climate Change on Water Resources in the West," Special issue, *Climatic Change* 62 (1–3) (January and February 2004).

"Phew, What a Scorcher—and It's Going to Get Worse." Agence France Presse, December 1, 2004. (Lexis).

Potter, Thomas D., and Bradley R. Colman, eds. *Handbook of Weather, Climate, and Water: Dynamics, Climate, Physical Meteorology, Weather Systems, and Measurements.* Hoboken, NJ: Wiley-Interscience, 2003.

Powell, Michael. "Northeast Seen Getting Balmier: Studies Forecast Altered Scenery, Coast." *Washington Post*, December 17, 2001, A-3.

"Quebec's Smoky Warning." Editorial, *Baltimore Sun*, July 9, 2002, 10-A.

"Rain, Rain Go Away." *London Times*, October 23, 2001. (Lexis).

"Record Rainfall Floods India." *New York Times*, July 28, 2005, A-12.

Reid, T. R. "As White Cliffs Are Crumbling, British Want Someone to Blame." *Washington Post*, May 13, 2001, A-23.

Revkin, Andrew C. "Forecast for a Warmer World: Deluge and Drought." *New York Times*, August 28, 2002, A-10.

———. "Global Warming Is Expected to Raise Hurricane Intensity." *New York Times*, September 30, 2004, A-20.

———. "2004 Was Fourth-Warmest Year Ever Recorded." *New York Times*, February 10, 2005. www.nytimes.com/2005/02/10/science/10warm.html.

Rubin, Josh. "Toronto's Blooming Warm: Gardens, Golfers Spring to Life as Record High Nears." *Toronto Star*, December 5, 2001, B-2.

Ryan, Siobhain. "National Icons Feel the Heat." *Australia Courier Mail*, February 4, 2002, 1.

Schar, Christoph, and Gerd Jendritzky. "Hot News from Summer 2003." *Nature* 432 (December 2, 2004): 559–561.

———, Pier Luigi Vidale, Daniel Luthi, Christoph Frei, Christian Haberli, Mark A. Linigier, et al. "The Role of Increasing Temperature Variability in European Summer Heatwaves." *Nature* 427 (January 22, 2004): 332–336.

Sevunts, Levon. "Prepare for More Freak Weather, Experts Say." *Montreal Gazette*, May 10, 2001, A-4.

Sharp, David. "Study: New England's Winters Not What They Used to Be." Associated Press State and Regional News Feed, July 23, 2003. (Lexis).

Simons, Paul. "Weatherwatch." *London Guardian*, November 26, 2001, 14.

Smith, Craig S. "One Hundred and Fifty Nations Start Groundwork for Global Warming Policies." *New York Times*, January 18, 2001, 7.

Spears, Tom. "Cold Spring Bucks the Trend." *Ottawa Citizen*, June 7, 2002, A-8.

Stewart, Fiona. "Climate Change in the Back Garden." *The Scotsman*, November 20, 2002, 8.

Stott, Peter A., D. A. Stone, and M. R. Allen. "Human Contribution to the European Heatwave of 2003." *Nature* 432 (December 2, 2004):610–613.

———, S. F. B. Tett, G. S. Jones, M. R. Allen, J. F. B. Mitchell, and G. J. Jenkins. "External Control of 20th Century Temperature by Natural and Anthropogenic Forcings." *Science* 290 (December 15, 2000): 2133–2137.

Struzik, Ed. "Fiery Future in Store for Forests If Climate Warms: Western Landscape Vulnerable: Study." *Edmonton Journal*, March 16, 2003, A-1.

"Summer Heat Wave in Europe Killed 35,000." United Press International, October 10, 2003. (Lexis).

Theobald, Steven. "Retailers Sweat It Out as Winter Sales Melt." *Toronto Star*, December 6, 2001, D-1.

Thompson, Dan. "Experts Say California Wildfires Could Worsen with Global Warming." Associated Press, November 12, 2003. (Lexis).

Thurman, Judith. "In Fashion: Broad Stripes and Bright Stars; The Spring-Summer Men's Fashion Shows in Milan and Paris." *The New Yorker*, July 28, 2003, 78–82.

Toner, Mike. "Temperatures Indicate More Global Warming." *Atlanta Journal and Constitution*, July 11, 2002, 12-A.

Townsend, Mark. "Monsoon Britain: As Storms Bombard Europe, Experts Say That What We Still Call 'Freak' Weather Could Soon Be the Norm." *London Observer*, August 11, 2002, 15.

Trenberth, Kevin E., Aiguo Dai, Roy M. Rassmussen, and David B. Parsons. "The Changing Character of Precipitation." *Bulletin of the American Meteorological Society* 84 (9) (September 2003): 1205–1217.

"2001 the Second Warmest Year on Record." Environment News Service, December 18, 2001. http://ens-news.com/ens/dec2001/2001L-12-18-01.html.

Uhlig, Robert, "Mild Autumn Produces England's First 'Noble Rot.'" *London Daily Telegraph*, December 6, 2001, Thursday, 1.

"U.S. N.S.F.: Scientists Find Climate Change Is Major Factor in Drought's Growing Reach." M2 Presswire, January 12, 2005. (Lexis).

Vaughan, David G., Gareth J. Marshall, William M. Connolley, John C. King, and Robert Mulvaney. "Climate Change: Devil in the Detail." *Science* 293 (September 7, 2001): 1777–1779.

Vergano, Dan. "Global Warming May Leave West in the Dust by 2050: Water Supplies Could Plummet 30 Percent, Climate Scientists Warn." *USA Today*, November 21, 2002, 9-D.

Vidal, John. "Better Get Used to It, Say Climate Experts." *London Guardian*, October 28, 2002, 3.

———. "The Darling Buds of February: Daffodils Flower and Frogs Spawn as Spring Gets Earlier and Earlier." *London Guardian*, February 23, 2002, 3.

Vogel, Nancy. "Less Snowfall Could Spell Big Problems for State." *Los Angeles Times*, June 11, 2001, A-1.

"Walnuts and Vineyards." *London Times*, October 29, 2001. (Lexis).

"Warmer Climate Could Disrupt Water Supplies." Environment News Service, December 20, 2001. http://ens-news.com/ens/dec2001/2001L-12-20-09.html.

Welch, Craig. "Global Warming Hitting Northwest Hard, Researchers Warn." *Seattle Times*, February 14, 2004. http://seattletimes.nwsource.com/html/localnews/2001857961_warming14m.html.

"White Christmases Just a Memory?" Environment News Service, December 21, 2001. http://ens-news.com/ens/dec2001/2001L-12-21-09.html.

"Why We're All Being Caught on the Hop by Global Warming." *Irish Independent*, July 17, 2004. (Lexis).

Wigley, T. M. L. "ENSO, Volcanoes, and Record Breaking Temperatures." *Geophysical Research Letters* 27 (2000): 4101–4104.

———, and S. C. B. Raper. "Interpretation of High Projections for Global-Mean Warming." *Science* 293 (July 20, 2001): 451–454.

Wilhite, Donald A. "Drought in the U.S. Great Plains." In *Handbook of Weather, Climate, and Water: Atmospheric Chemistry, Hydrology, and Societal Impacts*, ed. Thomas D. Potter and Bradley R. Colman, 743–58. Hoboken, NJ: Wiley Interscience, 2003.

Woods, Audrey. "English Gardens Disappearing in Global Warmth: Will Be Replaced by Palm Trees." Associated Press in *Canada Financial Post*, November 20, 2002, S-10.

Zeng, Ning. "Drought in the Sahel." *Science* 302 (November 7, 2003): 999–1000.

Zremski, Jeremy. "A Chilling Forecast on Global Warming." *Buffalo News*, August 8, 2002, A-1.

Zwiers, Francis W., and Andrew J. Weaver. "The Causes of 20th Century Warming." *Science* 290 (December 15, 2000): 2081–2083.